PROJECT MANAGEMENT

PROJECT MANAGEMENT

By
Prof. ABDUL MATHEEN
Department of Mechanical Engineering
H.K.B.K. College of Engineering
Bangalore

UNIVERSITY SCIENCE PRESS
(An Imprint of Laxmi Publications Pvt. Ltd.)
An ISO 9001:2015 Company
BENGALURU • CHENNAI • GUWAHATI • HYDERABAD
KOCHI • KOLKATA • LUCKNOW • RANCHI
NEW DELHI

PROJECT MANAGEMENT

Printed and bound in India
Typeset at : Goswami Associates, Delhi
Second Edition: 2009, *Reprint:* 2013, 2017, 2021; *Edition*: 2026
UPM-9355-459-PROJECT MANAGEMENT
Price: ₹ 459.00

Branches

Bengaluru	080-26 75 69 30
Chennai	044-24 34 47 26
Guwahati	0361-254 36 69
Hyderabad	040-27 55 53 83
Kochi	0484-405 13 03
Kolkata	033-40 04 77 79
Lucknow	0522-430 36 13
Ranchi	0651-224 24 64

PUBLISHED IN INDIA BY

UNIVERSITY SCIENCE PRESS
(An Imprint of Laxmi Publications Pvt. Ltd.)

An ISO 9001:2015 Company
113, GOLDEN HOUSE, GURUDWARA ROAD, DARYAGANJ,
NEW DELHI - 110002, INDIA

Corporate Office : B-33, Sector-5, Noida,
Gautambuddha Nagar (U.P.) - 201 301, INDIA
Telephone : +91-120-44 46 900
www.laxmipublications.com
info@laxmipublications.com

C—R/026/08
Printed at: Akashdeep Printer

CONTENTS

FOREWORD

Basics on project management usually abound with theoretical description and generally pose unnecessary difficulties for someone who's more concerned about managing project, eliminating wasting efforts by establishing closer links between project managers and down stream activities like development, manufacturing after sales support.

This book has unique features, that it takes a fresh look at the new techniques used in best practicing companies to improve their project management performance. This book enunciates principles and techniques and gives a direction that deeming approach to management can be applied to any kind of project in any industry.

The students will find the mixture of academic debate and practical case studies helpful and teachers will welcome the approach used in this text.

I believe that this is the essential virtue of the book. I recommend it not only to the enthusiastic students but also practicing managers in the field who many find it much useful in evolving new techniques in project management to eliminating waste time and effort in their system and make their system more efficient.

The major steps in project management are discussed in this book and help the managers to identify the constraints in the organisation and device more effective practice for best results.

Throughout the book, extensive examples are illustrated which helps any project manager to formulate policies for effective management.

PREFACE

Project Management is no longer about managing the sequence required to complete the project in time. It is about systematically incorporating the voice of the customer, creating a disciplined way of prioritizing efforts and resolving trade-offs, working concurrently on all aspects of the project in multifunctional teams. It also involves closer links between project team and down stream activities.

Project management is currently not over burdened with theory, but is practically based on the new management agenda which has been set up by the end of taylorism and the rise of participative management, the popular adoption of quality management and the empowerment of customers. The subject of management has grown from an off-shot of investigations into human behaviour to become a vast discipline. It has a technical specialism, not only of the area in which the project is taking place, but of the techniques associated with project management.

This book is addressed to the undergraduate students of various Universities who wish to understand the contributions made by Project Management, particularly final year students of Mechanical, Industrial Engineering and Management, and Industrial Production of **Visveswariah Technological University, Karnataka.** This book is based mainly on the lecture notes prepared for classes on the subject. I am grateful to all the authors of the books, I have used in preparing the notes.

I owe a debt of gratitude to the Honorable Chairman and Founder President of H.K.B.K. College of Engg. **Mr. C.M. Ibrahim (Former Union Minister, Govt. of India)** for the encouragement and constant support extended during this venture.

I express my sincere gratitudes to our beloved **Administrator Mr. Abdul Hameed. S.A.** and **Prof. Zaheer Ahmed, Principal H.K.B.K.C.E.,** for their motivation, guidance, cooperation and constant encouragement throughout the work on this book.

In the course of writing this book, I have received much encouragement from my colleagues, **Dr. J.F. Rahman,** H.O.D. (Mech.). Prof. R. Noor Ahmed, Prof. Tajuddin Yezdani, Prof. Wasee Mohd., Prof. P.F. Sanaulla, Mr. Abu-Sayeed, Mr. Mudassir Suhail Ghouri and Mr. Ayub Ahmed. Their efforts and labour of love under tight deadlines provided me extremely useful comments and encouragement ; I express my sincere gratitude to them. I am thankful to all my other colleagues, friends and students who encouraged me to write this book.

Despite my sincere efforts in checking the manuscripts and proof readings, mistakes might have crept in, I would very much appreciate for the detection of errors and suggestions for the improvement, from the readers of this book.

Author

1 CONCEPT OF PROJECT MANAGEMENT

INTRODUCTION TO PROJECT MANAGEMENT

Management system in today's world is filled with multidimensional challenges at various levels. The majority of corporate problems are based on High profitability, Low risk, effective utilization of existing resources, better control and use of manpower. Project management is one of the important internal solution.

Project management in the past, was confined only to the Department of Defense and Construction. Today the concept of project management is adopted in almost all the organizations such as Defense, Construction, Computer Software development, BPO, Mechanical Installation, O & M of Central Utilities, Hospital, Pharmaceuticals, Chemical, accounting, Banking, Advertisement, Law, State and Central Government Organizations. The project management approach requires a departure from the traditional vertical organizational system, which emphasizes a strong Superior-Subordinate relationship and is highly bureaucratic, which cannot respond rapidly to the changing environment.

Project management is no longer about managing the sequence required to complete the project in time. It is about systematically incorporating the voice of the customer, creating a disciplined way of prioritizing efforts and resolving trade-off, it involves closer link between project team and down stream activities, it empowers each and every team members to such an extent that they feel responsible not only for their own work, but for the entire team.

In project management, the empowered project team performs all activities in a participatory fashion ensuring that the diverse view points are considered with collective wisdom. The empowered team understand that greater empowerment means greater responsibility and they feel pride and passion towards their own work, they become quality conscious, self disciplined, high performers who add value to the project, the customer and the organization.

Definition of a Project

- A project is a 'non-repetitive activity'.
- It is goal oriented—it is being persued with a particular end or goal in mind;
- It has a particular set of constraints—usually centred around time and resource;
- The output of the project is measurable.

Definition of Management

- Management is the process of getting things done, by working both with and through people, operating in organised groups. Emphasis is placed up on the end objectives.
- Management is the development of a system model, characterized by input, processing, and output, and directly identifies the flow of resources necessary to achieve the set objectives.
- Management is a rational approach to decision making using a system of mathematical models and processes.
- Management is considered to be a system of cultural relationships involving social change.
- Management capabilities can be developed by studying the expensions of other managers whether or not the situations are similar.

DEFINITION OF PROJECT MANAGEMENT

Project management is the art of creating the illusion.

Project management includes planning, organising, directing and controlling activities in addition to motivating what is usually the most expensive resource on the project. The project planning involves deciding what has to be done, when and by whom. The resources then need to be organised through activities, such as procurement and recruitment. Directing their activities towards a coherent objective is a major management role. The activities also need controlling to ensure that they fit within the limits (*e.g.*, financial). Set for them.

The standard project management. Methodology for government information technology departments—defines a project in terms of its products.

These are catagorised as

- **Management.** The planning, documentation and control action of management.
- **Technical.** The planning, documentation and review of technical aspects of the project.
- **Quality.** The planning, documentation and review of the quality control of the project system.

PRINCE (Project in Controlled Environment)

Prince is a project management shell or structure within which plans can be formulated and actions controlled throughout the project life-cycle. Its major benefit is providing a degree of methodology standardisation between projects. This allows managers to concentrate on details of their specific project, confident in a recognised and proven method.

Applied to the project environment, the three—D's of project management describes the three phases. These are:

- **Design it.** Identify the need that the project will serve, construct models to show how the needs will be developed, evaluate these to determine the optimum process for the task and minimise risk;
- **Do it.** Carry out the project in line within the models or plans generated above;
- **Develop it.** Improve the models and processes in the light of the experience gained from the project, incorporating the check/study and act parts of the deming cycle.

There is no "most important phase". They represent a chain of activities. Like a chain, the project performance will only be as good as the weakest part of the process.

The last phase has a significant input to the performance of subsequent projects, in addition to elements of it determining the performance of the current project. There are three phases. Their application is summarised in Table 1.1.

Table 1.1 The Three Phases of Project Management

Phase	Key Issues	Fundamental Questions
Design it	Project and organisational strategy, goal definition, modelling and planning, estimating, resources analysis, conflict resolution and justification.	What is to be done ? Why is to be done ? How will it be done ? Who will be involved in each part ? When can it start and finish ?
Do it	Organisation, control Leadership-decision-making and problem-solving.	How should the project be managed on a day-to-day basis ?
Develop it	Assessment of process and outcomes of the project, evaluation, changes for the future.	How can the 'management process' be continually improved ?

There are a number of tasks and issues to be addressed in each phase. This provides a degree of complexity for the project manager and is one reason that there are few truly excellent examples of project management available.

Taking the analogy of the project as a chain, it is important that there is general competence across the phases. This is preferable to their being excellence in one area, with other areas falling down.

The generic life cycle for a project involves consideration of how the level of activity varies with time. This is illustrated in Fig. 1.1 and shows how the level of activity is relatively low during the design phase, increases through the doing phase, and decreases, through the development phase.

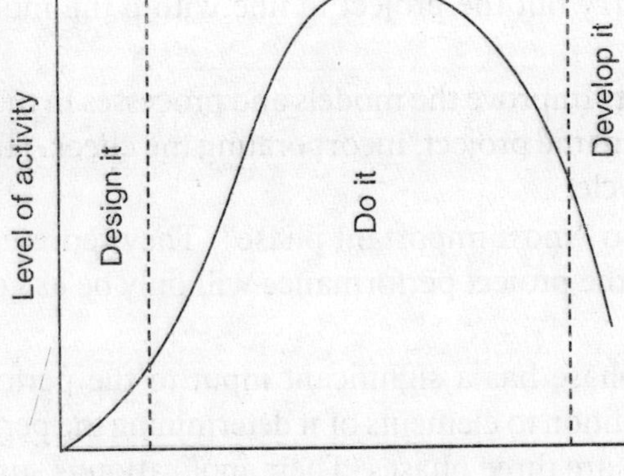

Fig. 1.1. Graph showing how level of activity varies with time.

This pattern is reflected in the graph of cumulative expenditure against time (Fig. 1.2) outgoings are generally low in the early stages, but grow rapidly during the execution phase.

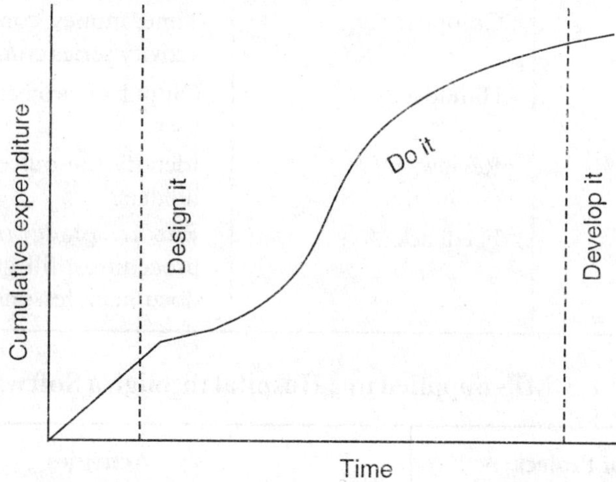

Fig. 1.2. Graph of cumulative expenditure against time.

The graph also demonstrates why the **"develop it"** phase is so vital by the time the majority of the doing phase is completed, the probability is that in excess of 98 per cent of the total project expenditure will have been incurred. The last phase is the time when the project team themselves can benefit from the process and ensure that lessons (good and bad) are applied in the future.

The life cycle may be further elaborated as shown in Table 1.2.

Table 1.2 Development of the Project Life Cycle

Stage in Project Life Cycle	Activity	Description
Design it	Conceptualisation	Generate explicit statement of needs.
	Analysis	Identify what has to be provided to meet those needs.
	Proposal	Show how those needs will be met through the project activities.
	Justification	Prepare and evaluate financial costs and benefits from the project.
	Agreement	Point at which go-ahead is agreed by proper sponsor.

Do it	Start-up	Gathering of resources, assemble project teams.
	Execution	Carry out defined activities.
	Completion	Time/money constraint reached or activity series completed.
	Handover	Output of project passed to client/user.
Develop it	Review	Identify the outcomes for all shareholders.
	Feedback	Put in place improvements to procedures, fill gaps in knowledge, document, lessons for the future.

Table 1.3 MIS Supplied to a Hospital through a Software Company

Phase of Project	Activities
Conceptualisation:	Software house receives an outline from the MIS department of the hospital; various pieces of information and points of classification are requested.
Analysis:	The concept is converted into the technology of the software house (every organisation has its own set of Jargon). An initial feasibility check is carried out to see what could be achieved at what cost. Objectives are set for the system to be developed and the interfaces with other system studied. The analysis phase was completed by an appraisal of the capability of the company to provide what was being asked for by the client.
Proposal:	The proposal document is submitted for approval by the client's MIS department in terms of whether or not it would meet the requirements set out in the initial request. The client organisation is offered the opportunity to visit the software house's premises and existing clients to view their systems.
Justification:	There are two parts to this process—first the software house doing their financial analysis to show whether or not it is feasible for them to undertake the project. The MIS people at the hospital need to provide evidence that the new system will provide a return. This has to be agreed by the financial managers.

Agreement:	After the justification has been prepared by both sides, the formal act of preparation and signing of contracts can take place. This is the basis of the agreement between supplier and customer. The terms and conditions will have to go through each of the party's legal advisers.
Start-up:	The software house starts to gather resources as soon as the contract looks likely to go ahead. Formal commitments are not made until the deal is formally signed. A project manager within the company is allocated to provide a single point of contact for the customer. The project team is gathered, external programmers hired and resources (development computers, pre-written software) procured. The project elements are allocated to individuals and specifications—written for what each of the elements must achieve.
Execution:	The project team starts work on the system—this is a mixture of importing existing code, modifying other parts and writing totally new elements. At the completion of each section of the work, the modules are tested to ensure integration. Gradually the system is pieced together, and de-bugged. The client is involved in the process, with modules being demonstrated as they are completed, so that amendments are made at the time, rather than at the end of the entire process.
Completion:	Towards the end of the development, the units being tested are getting larger and more complex. The in-house specialist staff are kept on and the programmers who were hired in continue to other jobs. The major task to be completed at this stage is the documentation of the system.
Handover:	The software is transported to the user's site and installed on the machines. The software specialists are on hand to see that any problems can be resolved quickly. Staff are trained in the usage of the system and the MIS staff on its maintenance and support. On going support is to be provided by the software house.

Check:	The way in which each of the modules was developed is documented to provide a rich picture of the process. Mistakes and good practice are identified, and customer perceptions of the system canvassed. The results to the company in financial terms are compared with the proposal.
Act:	Where deficiencies were highlighted *e.g.*, in the documentation of the system, the company put in place new procedures and practices that would ensure (*a*) that the problem for the customer was resolved in this case and (*b*) that it did not occur in the future.

Table 1.3 shows how a new management information system was supplied to a hospital through a software company.

Before we are ready to give the project any detailed consideration, further structure is needed. This prevents the project manager becoming lost in the morass of issues to be considered—particularly in more complex projects. The following 7-S framework allows segregation of the issues.

The 7-S of Project Management

The 7-S framework of management issues was promoted by McKinsey & Co., Management consultants. Their original 7-S is amended for the project environment, with a description of each of the elements, as shown in Table 1.4.

Table 1.4 The 7-S of Project Management

Element	Description
Strategy	The high-level requirements of the project and the means to achieve them.
Structure	The organisational arrangement that will be used to carry out the project.
Systems	The methods for work to be designed, monitored and controlled.
Staff	The selection, recruitment, management and leadership of those working on the project.
Skills	The managerial and technical tools available to the project manager and the staff.
Style/Culture	The underlying way of working and inter-relating within the work team or organisation.
Stake holders	Individuals and groups who have an interest in the project process or outcome.

Rather than being simply an outcome or a statement, strategy is a process. It involves a high-level consideration of objectives, which can be seen as points of principle, rather than activity-level details. Success starts with a rational strategy process, which then guides and informs the decision made in all areas of the project. The elements of strategy will be discussed further.

Along with the key issue of policy deployment—The manner in which organisational strategy is pursued through activities, including projects.

Structure is the arrangement of human resources relative to lines of command and control. A key question for the project manager concerns the nature of this structure. For example, should the project team be a dedicated, full-time team, or one where staff are 'borrowed' from other parts of the organisation or other organisations, only as and when needed ?

Systems are "the way we work". Both formal and informal systems will need to be designed or at least recognised for key tasks, including communication and quality assurance. Formal systems can be demonstrated through statements of procedure simply put, 'under there conditions, we carry out this action'—Informal systems, particularly for information transfer, are far less easy to describe and control. It is normal, however, for these to be the main mode of communication within groups. A theme within the systems elements is the focus of the systems on 'process'. That is, ensuring that all activities carried out are contributing to the end objective of the project in a constructive manner.

Staff need to be selected, recruited and then managed. How they respond to their treatment will have a large impact on the success or otherwise of the project. Yet this element has traditionally been neglected by texts on project management.

The 7-S framework provides a comprehensive set of issues that need to be considered. It also allows—classification of tasks within the limit of the project manager, which reduces the complexity of the role. In addition, classifying issues in this manner ensures that the project manager will know where to look to find sources of help if unusual—situations arise knowing that interpersonal problems in a team are aggravated by the style/culture that a project manager, promotes provides a means for finding solution to the problems.

The Project as a Conversion Process

The approach that will be considered in this book is a systems approach. The project is viewed as a conversion or transformation of some form of input into an output. As shown in Fig. 1.3, the inputs are some form of want or need which is satisfied through the process. The project will takes place under a set

which either provide the basis for any assumptions, or limit the project. The mechanisms are those resources that make the transformation process possible.

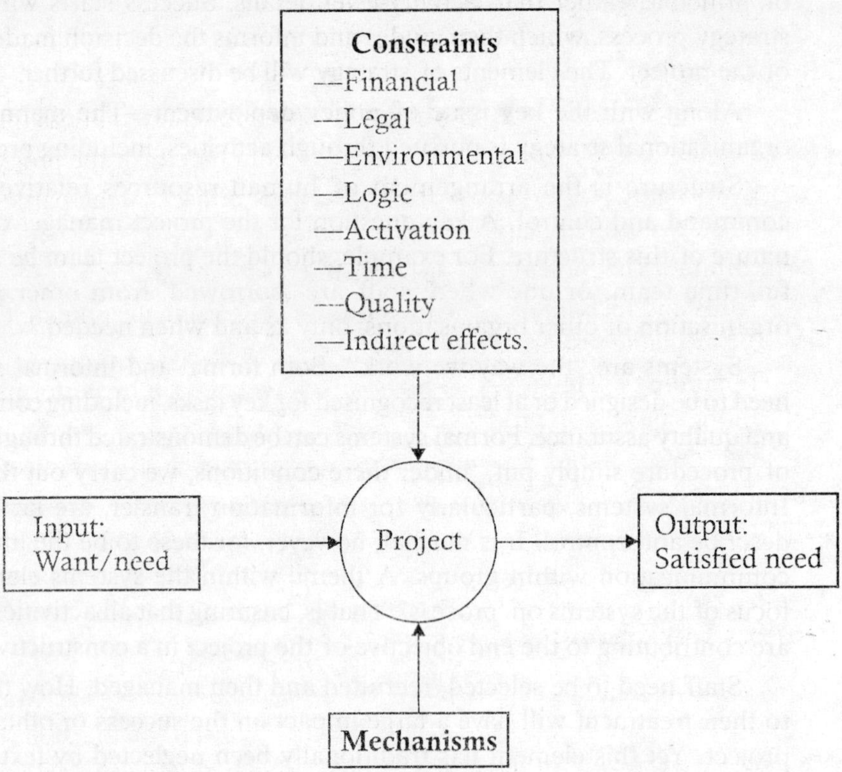

Fig. 1.3

The project as a coversion process:

—People
—Knowledge and Expertise
—Capital
—Tools and Techniques
—Technology.

Outputs

Fig. 1.3 describes the output as a 'Satisfied need'. This is a very wide interpretation of the possible outputs of a project and includes, for example, a new building as an output from a construction project; processed information, *e.g.*, in the form of engineering drawings or a report; and people with the necessary skills for a task (the output of a project involving training). The outputs may be tangible or intangible.

Mechanisms

The means or mechanism of which the output is achieved are as follow:

- **People.** Those involved both directly and indirectly in the project;
- **Knowledge and expertise.** Brought to the project by the participants and outside recruited help (*e.g.,* consultants);
- **Capital.** The money that provides the resources;
- **Tools and techniques.** The methods for organising the potential work with the available resources;
- **Technology.** The available physical assets that will be performing part of all the conversion process.

This consideration provides the most basic model of the project. It will be developed in Chapter 3 and 4 into more detailed consideration. The nature of the transformation process will be determined by the environment. In which it is taking place. This is the subject of the following section.

The Project Environment

There is the view that '*work can be done almost anywhere*'. The Mazda MXS (Miata) car, for example, was designed in California, financed in Tokyo and New York, tested in UK, assembled in Michigan and Mexico using components designed in America and Made in Japan. Ship building a predominantly project based environment has faced competition coming from parts of the World that 20 years ago had little or no capability in this area. Traditional not for profit organisations (including the National Health Service) are now required to meet performance targets and individual activities are being subjected to previously unthinkable commercial constraints.

Within the Expanding European Union, trade barriers have fallen and the number of international collaborative ventures has increased. Projects have become more complex as:

- Businesses are becoming more complex—it is less likely for a company to provide a commodity product or service, but to provide a 'package' which meets an entire need rather than just part of that need;
- Generally the simplest ideas have been exploited first—it is becoming more difficult but more vital to be innovative;

- Projects are moving towards turn-key contracts—where the end user does none of the interfacing between the different parts of the system, but deals with a single supplier in the provision of an entire system.

Concept of a Project

A project is defined as 'non-repetitive activity and is basically a planning procedure related to setting up of a business or accomplishing a particular activity to enhance the value of business. Every project has a different approach and the characteristics of a project are as follows:

1. Non-routine and non-repetitive undertaking often plagued with uncertainties.
2. Involves co-ordination of persons involved in different functional areas and contributions from different functional agencies and their efforts.
3. Setting up of dynamic, temporary and flexible relationship among people.

Need for a Project

In a business setting, whether in the public or private sector, an organisation must grow atleast for the sake of its survival. The organisation is hence on the outlook for good business ideas, which may require growth, either on existing lines of business or in diversified areas.

But the idea must be technically feasible and economically viable apart from being politically and socially acceptable. Once the feasibility has been justified an investment proposal is made. The approval of the investment proposal signifies the commencement of the project. A project is thus initiated to achieve a mission whatever the mission may be and is said to be completed when the mission is achieved. The project lives between these two cut off periods and this intermediate time is termed as **"Project Life Cycle"**.

The project management institute, U.S.A. defines project as a "*one short, time limited, goal depended undertaking, requiring the commitment of varied skills and resources*". The resources may be human or non-human, pooled together in a temporary organisation to achieve a specific purpose.

"Traditional Forms of Organisation are not Suitable for Projects" a Comment

Traditional management structures have been in place for over 200 years, but the recent trends in business has forced. Many companies to change their existing organisation structures.

The traditional or classical organisation is structured as shown below:

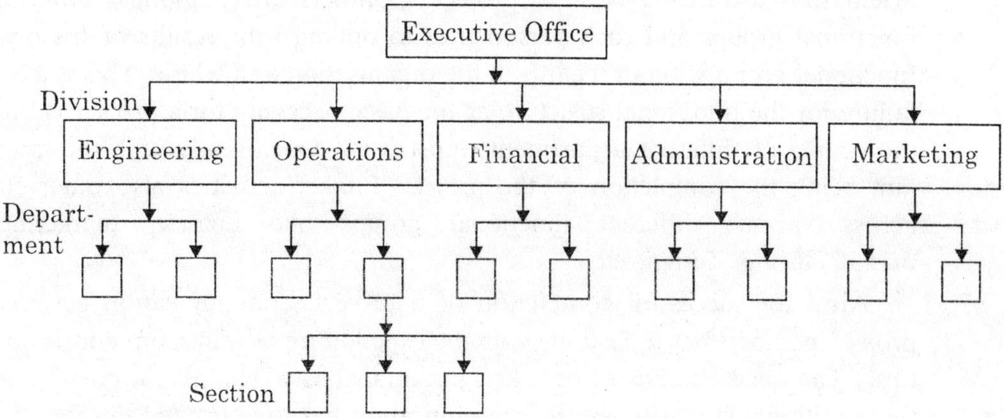

Each activity has its own department of functional group to do tasks related to that activity. Each department also maintains a strong technical expertise pertaining to that area. Some advantages of the classical organisational structure are:

1. Easier budgeting and cost control.
2. Better technical control as each department has experts.
3. Flexible use of manpower.
4. Policies, procedures and lines of responsibility are clearly defined.
5. Enables mass production activities that have well defined specifications.
6. Provides good communication channels and good control over personnel.

Inspite of the above advantages, the classical structure is not suitable for projects as its disadvantage out weigh its advantages. Some major disadvantags are:

1. No individual is responsible for the total project.
2. It does not provide the project-oriented emphasis necessary to accomplish project.
3. Co-ordination becomes complex leading to higher lead times.
4. Lack of a clear customer focus and response to customer is slow.
5. Decisions normally favour the strongest functional group.
6. Difficulty in pin pointing responsibility, as there is no direct project reporting and also due to poor project oriented planning.
7. Motivation and innovation is decreased.

The disadvantages basically arise due to the fact that there is no clear orientation towards a definite project. Conflicts arise amongst different functional groups and each person tries to optimise the results of his own functional group without regards to the organisation as a whole. This is a big failure for the traditional structure as the basic necessity for a project is that the people of different departments must work concurrently with the main aim being the completion of the project. Project activities also often cut across various different functional groups like finance, marketing, manufacturing, design, etc.

Also, for successful completion of a project, communication between project members has to be done with the expenditure of minimum efforts and time. The decisions has to be taken instantaneously. This is not possible in the traditional structure, as all communication has to be routed through the top management. Thus, destroying valuable time and delaying the decision process.

Customer focus is also an important part of projects, which cannot be achieved in the traditional structure, as the classical. Setup has no customer focal point.

The classical setup is basically suited for a system wherein mass production is needed. But a project is basically a one off or one time activity that is done over a fixed life span.

Due to the above reasons, projects have a tendency to fall behind schedule in the classical structure.

Thus, the traditional structure is not suitable for projects.

Characteristics of Project

1. Fixed set objectives. Once they are met, project is closed.
2. Project has a definite life span.
3. The project is a single entity, while its participants may be many.
4. Team work is absolutely essential. People from all functional groups are needed.
5. Projects have definite life cycles.
6. All projects are unique, as the people and plans involved in similar projects also are different.

7. A project is subjected to a lot of change.
8. Made to order.
9. Has a high level of sub contracting.
10. A project is fraught with risk and uncertainly.

Categories of Projects

Projects can be categorised as:

1. **Individual projects.** These are short duration projects normally assigned to a single individual who may be acting both as a project manager and as a functional manager.
2. **Special projects.** This require certain primary functions and/or authority to be assigned temporarily to other individuals. This arrangement can lead to severe conflicts in case of long term projects.
3. **Matrix or aggregate projects.** These type of projects require inputs from a large number of functional units and usually control vast resources.
4. **Staff projects.** These are accomplished by a single organisational unit or department. This is suitable if only one functional unit is involved.

Apart from the above categorisation, projects can also be categorised based on their location, type, technology, size, scope and speed. Categorisation is important because though all projects are similar in structure, they differ a lot when it comes to the amount of effort and resources necessary for the project. Thus managements must understand clearly the categorisation.

Based on the speed of execution of a project, they can be classified as:

1. **Normal projects.** Adequate time is provided for implementation. Project will involve minimum capital cost and no sacrifice in terms of quality.
2. **Crash projects.** Additional capital cost are incurred to gain time. In this case the quality may be compromised.
3. **Disaster projects.** Anything needed to gain time is allowed in this type of projects. Quality short of failure level is also accepted. Round the clock work is usually done.

The schematic diagram given below indicates the categories of projects.

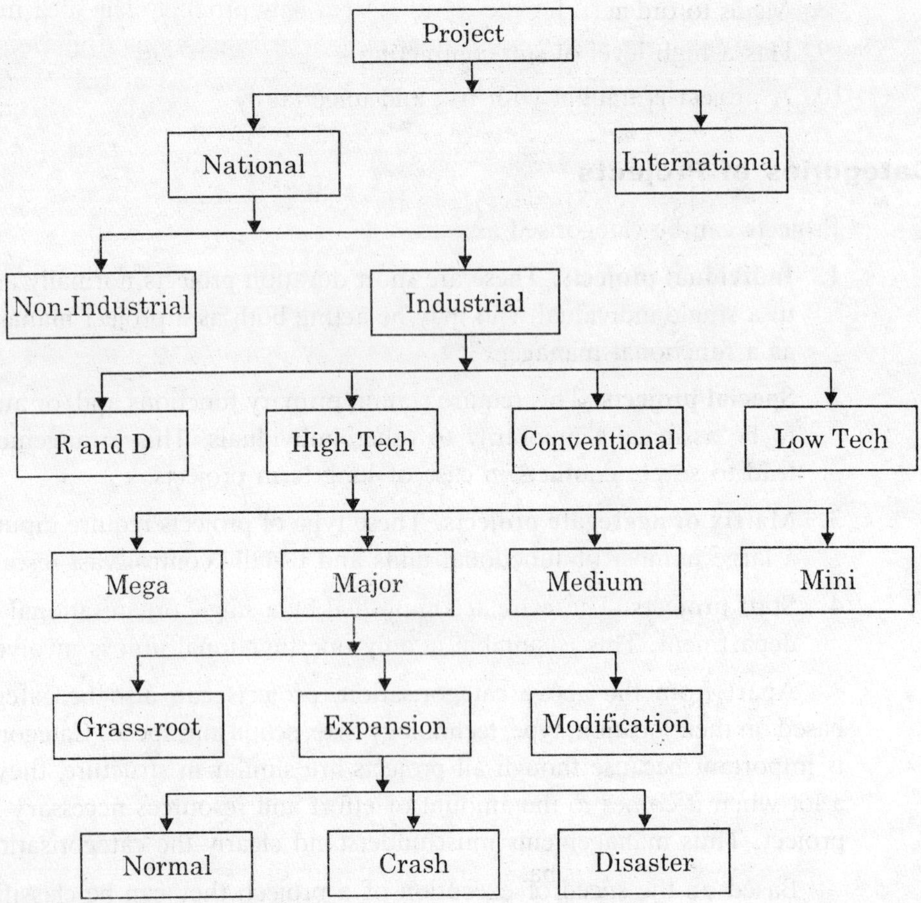

Phases of Project Life Cycle

1. Conception or conceptual phase.
2. Definition phase.
3. Planning and organising phase.
4. Operation or implementation phase or production phase.
5. Project clean up or divestment phase.

Conceptual Phase

This is the first phase and it includes the preliminary evaluation of the idea. The idea may first come to the mind when one is seriously trying to overcome—some problems.

The problems may be ranging from the underutilisation of plant capacity and funds to the need for the creation of a new product. The idea may also originate from the surrounding-environment or suggestions from people. An important task in this phase is the preliminary analysis of risk. The efforts attributed to this phase must include:

(*a*) Determining the existing needs or potential deficiencies of existing system.
(*b*) Establishing systems and concepts that provide guidance to overcome existing or potential deficiencies.
(*c*) Determine initial technical, economic and environmental feasibility.
(*d*) Examine alternate ways of achieving system objectives.
(*e*) Identify human and non-human resources required to support the system.
(*f*) Select initial system design that will satisfy the system objectives.
(*g*) Determine initial system interfaces and establish a system organisation.

Real Project Example

Consider a cement plant operating at low capacity utilisation, high power consumption and hence high operating costs. Some possible ideas that have to be thought or to improve the situation are:

- Introduce new technology.
- Replace critical parts selectively.
- Scrap the plant once for all.

Definition Phase

The definition phase is basically a refinement of the elements described in the previous phase. This phase requires a firm identification of the resources to be requires together with the establishment of realistic time, cost and performance parameters. This phase also involves the preparation of initial documentation necessary to support the system. For a project based upon competitive bidding, the conceptual phase would include the decision to bid or not while the definition phase would include the development of the total bid package. This phase must include the following activities.

(*i*) Preparation of final system performance requirements.
(*ii*) Preparation of detailed plans to support the system.
(*iii*) Determination of realistic cost, schedule and performance requirements.
(*iv*) Identification of those areas of the system where high risk and uncertainty is involved.

(*v*) Determination of necessary support systems.
(*vi*) Identification and initial preparation of documents required to support the system such as policies, procedures, budget, job descriptions, etc.

Real Project Example

The areas to be examined in the definition phase for a cement plant are:

- Nature and type of raw materials to be used.
- Plant size and capacity.
- Location and site.
- Technology and type of process to be used.
- Project layout.
- Plant and machinery.
- Electrical, civil and instrumentation work.
- Manpower and organisation pattern.
- Financial analysis and implementation schedule.

Planning Phase

This phase is predominantly a testing and final standardisation effort so that operations can begin. Almost all documentation must be complete in this phase. This phase must include the efforts such as:

1. Updating the detailed plans conceived and defined during the preceding phases.
2. Identification and management of resources required to facilitate the production processes such as inventory labour, etc.
3. Verification of system production specifications.
4. Beginning of production, construction and installation.
5. Final preparation and dissemination (distribution) of policy and procedural documents.
6. Performing final testing of the system to check if its performance matches the planned requirement.
7. Development of technical manuals and affiliated documents describing how the system is intended to operate.
8. Development of plans to support the system during its operation phase.

Implementation Phase/Operational Phase

This is a phase involving hectic activity as people can see the project moving towards completion. This phase is very important as bulk of the work is done in this phase. Implementation/operational phase is very similar to the definition phase and runs parallel to it to a certain extent. This phase must include:

1. Use of system results by the intended user or customer.
2. Actual integration of the project's product or service into existing organisational systems.
3. Evalution of technical, social and economic sufficiency of the project to meet actual operating conditions.
4. Provision of feedback to the organisation planners regarding the development of new projects and systems.
5. Evaluation of adequacy of support systems.

Divestment Phase

This phase includes the reallocation of resources and people to different projects as the existing project reaches its completion point. This phase evaluates the efforts on the trial system and serve as important to the conceptual phase for new projects and systems. It has an impact on other ongoing projects with regard to priority identification.

In this phase, the project accounts are closed, material reconciliation is done, outstanding payments made and dues collected. This phase includes:

1. System phase down.
2. Development of plans transferring responsibility to supporting systems.
3. Divestment or transfer of resources to other systems.
4. Listing the lessons learned and creation of a data base including:
   (*i*) Customer feedback.
   (*ii*) Major problems encountered and their solution.
   (*iii*) Technological advances.
   (*iv*) Recommendations for R and D.
   (*v*) New or improved management techniques.

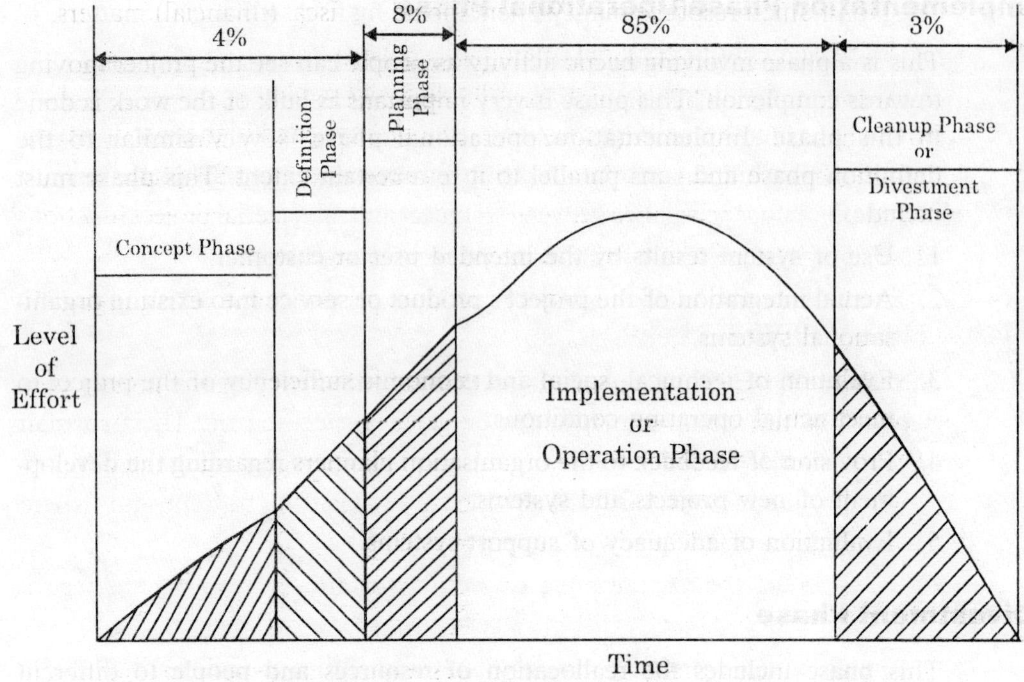

Fig. 1.4. Level of efforts *Vs* time for the project life.

Roles and Responsibilities of Project Manager

The roles and responsibilities of a project manager vary, depending upon the nature of the project and also depending upon the number of project managers in the project. But, by and large the roles of a project manager can be clearly spelt out considering the basic responsibilities of all project manager in any project. They are:

1. Co-ordinating and integrating activities across multiple functional lines.
2. Defining and maintaining the integrity of the project.
3. Development of project execution plan.
4. Organising the executed plan.
5. Setting targets and development of systems and procedures for accomplishment of project objectives and targets.
6. Negotiations for commitments from suppliers, clients and project members.
7. Direction, co-ordination and control of project activities.

8. Non-human resource management including fiscal (financial) matters.
9. Management of personnel (man management).
10. Satisfy government, customer, promoters and public.
11. Achievement of cash surplus—project objectives and higher productivity.
12. Managing human inter-relationships in the project organisation.
13. Maintaining the balance between technical and managerial project functions.
14. Coping with risk associated with project management.

Functional Manager's Role

Assuming that the project manager and the functional manager is not the same person, the functional manager can be given a specific role. The two main elements to his role are:

1. He has the responsibility to define 'how' the task will be done and 'where' it will be done.
2. He also has the responsibility of providing sufficient resources to accomplish the objective with the project's constraints.

In other words, once the project manager identifies the requirement for the project, it becomes the line manager's responsibility to identify the technical criteria. The only exception is in the area of R and D where the line manager is usually more technically sound. The line manager also has the duty and powers of allocating resources and appointing jobs to different people. The functional or line manager is usually under tremendous pressure to live up to their commitments. They have to cope with:

1. Unlimited work requests.
2. Predetermined deadlines.
3. Complying requests from higher authorities.
4. Limited number and availability of resources.
5. Unscheduled changes in project plan.
6. Unplanned loss, break down or absence of resources.
7. Unpredicated lack of progress.

Thus in conclusion it can be said that, the project managers plan, monitor and control the projects whereas the functional managers are responsible for performing the work.

Tools and Techniques

Project management has a special set of tools and techniques, which has scientific approaches enabling the project managers to plan and evaluate the progress of the project. But, these tools and techniques are not an end in themselves as they are capable of generating only numbers and not solutions. It is left to the human wisdom, to interpret these numeric results and implement changes in the project suitably. Thus, the tools indicate 'what' is to be done, but do not indicate 'how' it must be done.

The Tools and techniques can be broadly grouped as:

1. **Project Selection Techniques:**
   (*a*) Cost benefit analysis
   (*b*) Risk and sensitivity analysis.
2. **Project Execution Planning Techniques:**
   (*a*) Work Breakdown Structure (WBS)
   (*b*) Project Execution Plan (PEP)
   (*c*) Project Responsibility Matrix
   (*d*) Project Management Manual.
3. **Project Scheduling and Co-ordinating Techniques:**
   (*a*) Bar charts
   (*b*) Line of balance
   (*c*) Network diagram (PERT/CPM)
   (*d*) Life cycle curves.
4. **Project Monitoring and Progressing Techniques:**
   (*a*) Progress Measurement Technique (PROMPT)
   (*b*) Performance Monitoring Technique (PERMIT)
   (*c*) Updating, Reviewing and Reporting Technique (URT).
5. **Project Cost and Productivity Control Technique:**
   (*a*) Value engineering
   (*b*) Productivity budgeting technique
   (*c*) Cost/WBS.
6. **Project Communication and Clean-up Technique:**
   (*a*) Control room
   (*b*) Computerised information system.

Review Questions

1. Define Project Management.
2. What do you understand by "PRINCE"? Enumerate the various phases involved in project environment with an example.
3. Explain the various elements of Project Management.
4. Justify the statement "Project is a Conversion Process".
5. "Traditional forms of organisations are not suitable for project". Comment.
6. Briefly explain the characteristics of a Project. State how projects can be categorised?
7. What do you understand by the term "Life Cycle"? Illustrate different phases of project life cycle with example.
8. Explain the roles and responsibilities of Project Manager.
9. Explain the role of Functional Manger.
10. Explain the tools and techniques of Project Management.

2 PROJECT PLANNING AND ESTIMATION

FEASIBILITY REPORT

The important facts of project analysis (Feasibility Report) are:

1. Market feasibility
2. Technical feasibility
3. Financial feasibility
4. Economic feasibility
5. Ecological feasibility.

Market feasibility

This is concerned with the aggregate demand of the concerned product and level of market share under appraisal. The market analysis requires a wide variety of information and appropriate forecasting method related to:

(*a*) Consumption trends in the past and present level.

(*b*) Past and present supply position.

(*c*) Production possibilities and constraints.

(*d*) Imports and exports.

(*e*) Structure of competition.

(*f*) Cost structure.

(*g*) Elasticity of demand.
(*h*) Consumer, behaviour, intentions, motivations, attitudes, preferences and requirements.
(*i*) Distribution channels and marketing policies in use.
(*j*) Administrative, technical and legal constraints.

Technical Feasibility

Analysis of the technical and engineering aspects of a project need to be done continually when a project is formulated. Technical analysis seeks to be determine whether the pre-requisites for the successful commissioning of the project have been considered and reasonably good choices have been made with respect to location, size, process, etc.

The important questions raised during technical analysis are:

(*a*) Whether the preliminary tests and studies have been done?
(*b*) Whether the availability of raw materials, power and other inputs have been established?
(*c*) Whether the selected scale of operation is optional?
(*d*) Whether the production process chosen is suitable?
(*e*) Whether the equipments and machines chosen are appropriate?
(*f*) Whether the auxiliary equipments and supplementary engineering works have been provided for?
(*g*) Whether the provision has been made for the treatment of effluents?
(*h*) Whether the proposed site layout, building, plant is sound?
(*i*) Whether work schedules have been realistically drawn up?
(*j*) Whether the technology proposed to be employed is appropriate from the social point of view?

Financial Feasibility/Viability

Financial analysis seeks to ascertain whether the proposed project will be financially viable in the sense of being able to meet the burden of servicing debt and whether the proposed project will satisfy the return expectations of those who provide the capital. The aspects that have been looked into while conducting the financial appraisal are:

1. Investment outlay and cost of project.
2. Means of financing.

3. Cost of capital.
4. Projected profitability.
5. Break-even point.
6. Cash flows of the project.
7. Investment worthwhileness judged in terms of various merit criteria.
8. Projected financial position.
9. Level of risk.

Economic Feasibility/Viability

Economic feasibility is also referred to as the socio-economic benefits where the cost of project is concerned with judging a project from a larger social point of view. The evaluation is focussed on the social costs and benefits that may often differ from its monetary costs. The social cost benefit analysis focuses towards:

1. The impact of the project on the distribution of income in the society.
2. Impact of the project on the level of saving and investment in the society.
3. Contribution of the project towards fulfilment of certain merits like self sufficiency, employment and social contribution.
4. Direct economic benefits and costs of the project measured in terms of shadow prices and not market prices.

Ecological Feasibility/Viability

In recent years, environmental concerns have assumed a great deal of significance and rightly so. Ecological analysis should be done, particularly for major projects which have significant ecological implications like power plants, irrigation schemes and environmental pollution industries like bulk drugs, chemicals and leather processing.

The key questions raised in ecological analysis are:

(*a*) What is the damage likely caused by the project to the environment?

(*b*) What is the cost of restoration measures required to ensure that the damages to the environment is contained within acceptable limits?

The schematic diagram given below indicates the feasibility study:

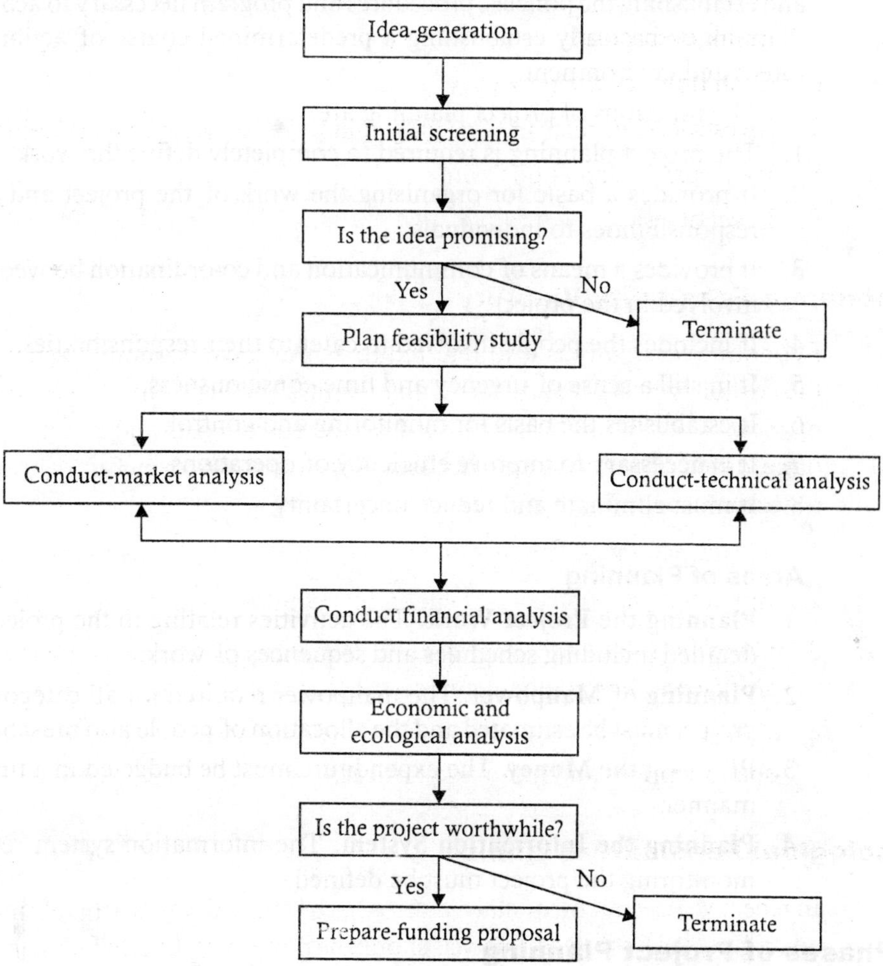

Fig. 2.1

Project Planning

The most important responsibilities of a project manager are planning, integrating and executing plans. Almost all projects, because of their relatively short duration and often prioritised resources, require formal and detailed planning. The integration of planning activities is essential to co-ordinate the individual plans of the various functional groups.

Planning can be described as the function of selecting the enterprise objectives and establishing the policies, procedures and program necessary to achieve them. Planning is basically establishing a predetermined coarse of action within a forecasted environment.

The functions of project planning are:

1. The project planning is required to completely define the work.
2. It provides a basic for organising the work of the project and allocating responsibilities to individuals.
3. It provides a means of communication and co-ordination between all those involved in the project.
4. It includes the people involved to cater to their responsibilities.
5. It in still a sense of urgency and time consciousness.
6. It establishes the basis for monitoring and control.
7. It is necessary to improve efficiency of operations.
8. It must eliminate and reduce uncertainty.

Areas of Planning

1. **Planning the Project Work.** The activities relating to the project must be detailed including schedules and sequences of work.
2. **Planning of Manpower.** The manpower required for all categories of the project must be estimated and the allocation of people also must be planned.
3. **Pl ng the Money.** The expenditure must be budgeted in a time phased manner.
4. **Planning the Information System.** The information system required for monitoring the project must be defined.

Phases of Project Planning

The phases of project planning can be classified as:

(*a*) Pre investment phase

(*b*) Implementation phase

(*c*) Operation phase.

(a) The Planning Activities in the Pre Investment Phase:

1. Estimating the demand and supply patterns of the product.
2. Establishing the import and exports over a period of time.
3. Determining the scope for expansion from existing product line.

4. Identifying the consumption patterns in other locations.
5. Determining the natural resources needed for the production activity.

Once the investment opportunity has been seen, preliminary project analysis is done that involves determining the marketing, technical, financial, economic and ecological feasibility. The analysis of the feasibility studies more or less enable the establishment of location, production capacity, production technology and other inputs.

(b) The Planning Activities in the Implementation Phase

The implementation phase for an industrial project involves setting up of manufacturing facilities consisting of several stages namely.

1. **Project engineering and designs.** This must include preparation of construction drawings and blueprints, plant designs, plant engineerings, selection of specific machinery and equipment, determining water and power requirements, safety devices, etc.
2. **Negotiations and contracting.** This phase includes drawing up of legal contracts with respect to project financing and also contracting outwork like construction of buildings, supply of machines and equipments, civil works, etc.
3. **Construction.** This includes the actual construction of plants and erection of machineries and equipments.
4. **Training.** This includes the training of engineers, technicians and workers.
5. **Plant commissioning.** This deal with planning the start-up of the plant.

(c) The Planning Activities in the Operation Phase

This is the longest phase in terms of time span. It begins when the project is commissioned and ends when the project is wound up. The planning activities in this stage must be aimed at enabling. Uninterrupted and smooth functioning of the plants and machines. This planning must also include developing some norms of productivity.

Five Steps of Project Management

1. Group work into packages that acquires the properties of the project. The group works are related to each other but contribute to the same goal.
2. Entrust the whole project to a single responsibility that is the project manager for co-ordinating, directing and controlling the whole project.

3. Support the project internally by having a matrix type of organisation and support it externally through vendors and contractors.
4. Build up commitment by directing towards the goals through schedules and costs.
5. Ensure adherence to goals by constant monitoring.

Objectives and Goals of Project Planning

Objectives and goals of project planning depends upon the product produced, type of industry and location of industry.

The objectives can be classified as:

1. Performance objectives.
2. Technical objectives.

Performance Objectives

A project requires a systemic performance analysis based on its objectives, namely:

(*a*) Determining if the project as a whole is on schedule, ahead of schedule or behind schedule.

(*b*) If any failures occur, who is responsible and what are the implications.

(*c*) If the cost of the projects is a whole or in part is as per budgeted requirements.

(*d*) If variations did occur, what were the reasons and where did it occur.

The performance should be evaluated using various tools and techniques on scheduling, allocation skills and cost controls. For small projects, the objectives may be based on simple analysis, but for large projects, the objectives must be defined clearly by top management.

Technical Objectives

These are the objectives based on the technical analysis and engineering aspects of the projects. The technical objectives are concerned with:

(*a*) Material inputs and utilities.
(*b*) Product mix.
(*c*) Plant location and site.
(*d*) Plant capacity.
(*e*) Machine and equipment.
(*f*) Structural and civil works.

(*g*) Works scheduling techniques and quality standards.

(*h*) Market analysis.

Financial Arrangements

For the zero date of the project to become effective, it is essential that funds required for the project are arranged. The fund requirement for a project is to meet the capital expenditure for purchasing plant and machinery, initial working capital and pre-operative expenses through the entire fund is not needed on zero date, suitable arrangement has to be made in advance so that funds do not pose a constraint for meeting the project targets.

Some important concepts in financing arrangement are:

Capital Costs

All the cost incurred in the project before it becomes ready to start commercial production will be treated as capital costs.

Capital costs not only include expenditure on assets like land, plant, machinery, etc., but also on software costs, pre-operative expenses, license fee, etc.

Working Capital

The fund required for maintaining various inventories in the form of raw materials, operating supplies, intermediate products, finishing products and meeting miscellaneous cash requirement is treated as working capital. The working capital is that is borrowed by hypothecating raw materials is termed as borrowed working capital. The rest of the initial working capital financed by the promoter himself is termed as margin money usually margin money must be 20% to 30% of initial requirement.

Operating Costs

The costs that are incurred on a recurring basic for production, maintenance and marketing are termed as operating costs. Raw materials, utilities, interest on loans, selling expenses and any other expenses incurred year after year once the project has entered commercial production is treated as operating expenses.

Sources of Financing

The three sources of financing are:

Table 2.1

Term of Financing	Source	Expenditure to be Financed
1. Short term financing	• Trade credit. • Loan from commercial banks • Accounts receivable.	Working capital
2. Intermediate term financing	Term loans Hire purchase Lease financing Fixed deposits from public.	Capital cost
3. Long term financing	Equity stock debt preferred stock.	Capital cost

Preparation of Cost Estimates

A project cost estimate is required not only for assessing fund requirement but also for ascertaining the economic viability of the project. The estimate should be accurate. If the cost of the project is not estimated accurately, the fund plan will consequently be inaccurate. The unit in such a case will face an acute shortage of funds and this may lead to the suspension of the project or at least a considerable delay. This in turn increases overall project cost rendering it economically viable or sick.

On the other hand, if the project cost is overestimated the project may fail to receive the green signal due to its lower profitability in comparison to other competitive projects. If excessive funds are committed to any project due to overestimate, several other needy projects will be unnecessarily deprived of funds. An accurate estimate is thus necessarily not only for getting a project cleared but also to ensure adequate profitability.

But accuracy in a project is not easy to ensure because of the non-availability of the required information. The information gets developed only as the project progresses, but the cost estimate is required on the first day itself so as to assess the magnitude of the investment involved.

Thus, estimates have to be made at each stage of the project life cycle so that a working estimate of the cost is obtained that has a reasonable accuracy.

EVALUATION OF PROJECT PROFITABILITY

The matter given below indicates how (the methods) the economic viability or project profitability is determined.

At the zero date, both the cost estimate and the overall schedule would have changed from what they were in the feasibility report. This changes the profitability of the project changes continue as the project progresses, but despite these changes, the project must still remain profitable or economically viable. There can be no questioning of proceeding with the project if the project is not economically viable.

Methods used to assess economic viability are:

1. Pay Back Period (PBP)
2. Return on Investment (ROI)
3. Net Present Value (NPV)
4. Internal Rate of Return (IRR)
5. Social Profitability.

Pay Back Period (PBP)

This is the time required to recover the original investment through incomes from the project.

$$\text{PBP} = \frac{\text{Original investment (Rs.)}}{\text{Annual income (Rs.)}} = \text{Number of years.}$$

PBP essentially reflects how soon the money sunk in the project is going to be recovered.

PBP is used in projects financed with high borrowed money, projects with high risk and projects where technological obsolescence is very rapid.

However, PBP does not take the whole life of the project into account and also the information on total income. Thus, this method neither indicates the profitability of investment nor the total worth of the project.

Return on Investment (ROI)

The ratio relates earnings to investments

$$\text{ROI} = \frac{\text{Average annual earning after tax}}{\text{Average book investment after depreciation}}.$$

E.g.

Let capital investment	= Rs. 150 lakh
Project operating life	= 10 years
Salvage value after 10 years	= 0
Expected profit after tax	= Rs. 15 lakhs for first 5 years and Rs. 7.5 lakh for the balance 5 years.

$$\text{ROI} = \frac{(15 \times 5 + 7.5 \times 5)/10}{(150 + 0)/2} \times 100 = 15\%$$

ROI must not only be higher than basic borrowing rate, but also above the minimum acceptable profitability rate of the company. This method is better than PBP as it emphasis on profitability and not liquidity of investment. It takes into account the whole project life and total income accurable to a project during this period.

This is useful for projects with short operating lives.

Net Present Value (NPV)

The above two methods do not discuss the time value of money. It can be easily appreciated that a rupee today has a higher value than a rupee after a year from now. This is because of the interest that the rupee is going to earn over the next one year period.

NPV for any project is the aggregate present value of the net cash flows over the operating life of the project.

$$\text{NPV} = \sum_{t=0}^{n} S_t \frac{1}{(1+r)^t}$$

where, NPV = Net present value

S_t = Net cash flow for the year

t = Operating year

n = Operating life of project

r = Interest rate.

The cash flow when $t = 0$ is the initial investment. This is considered as outflow and is treated as negative cash flow for computation purposes. All incomes are possible and known as cash inflow. A project must have a positive NPV. A project with higher NPV is better.

Internal Rate of Return (IRR)

This is again a discounting method, but in this case instead of assuming a fixed discount rate, the discount rate is varied till the net present value becomes zero.

The discount rate at which the net present value becomes zero is known as internal rate of return. IRR then reflects the true yield of a project and should enable comparison of the project with others. Project with high IRR are considered the best.

Social Profitability

The methods described so far have examined the financial yields from the project. While the financial yields are important for most industrial projects, the socio-economic yields also are equally important. The social profitability analysis must include studying the factors such as foreign exchange savings, development of backward areas, self reliance defence requirement. This type of analysis is found more commonly in public sector companies where making profits is not the only relevant consideration.

Financial Analysis

To judge a project from the financial angle, we need information about the following:

1. Cost of project.
2. Means of financing.
3. Estimates of sales and production.
4. Cost of production.
5. Working capital requirement and its financing.
6. Estimates of working results.
7. Break-even-point.
8. Projected cash flow statements.
9. Projected balance sheets.

This chapter discusses the manner in which the above information, essential for proper financial evaluation, is gathered, prepared, summarised, and presented. Before we dwell on the above financial estimates and projections, it is helpful to know how they are inter-related.

Cost of Project

Conceptually the cost of project represents the total of all items of outlay associated with a project which are supported by long-term funds. It is the sum of the outlays on the following:

1. Land and site development.

2. Buildings and civil works.
3. Plant and machinery.
4. Technical know-how and engineering fees.
5. Expenses on foreign technicians and training of Indian technicians abroad.
6. Miscellaneous fixed assets.
7. Preliminary and capital issue expenses.
8. Pre-operative expenses.
9. Provision for contingencies.
10. Margin money for working capital.
11. Initial cash losses.

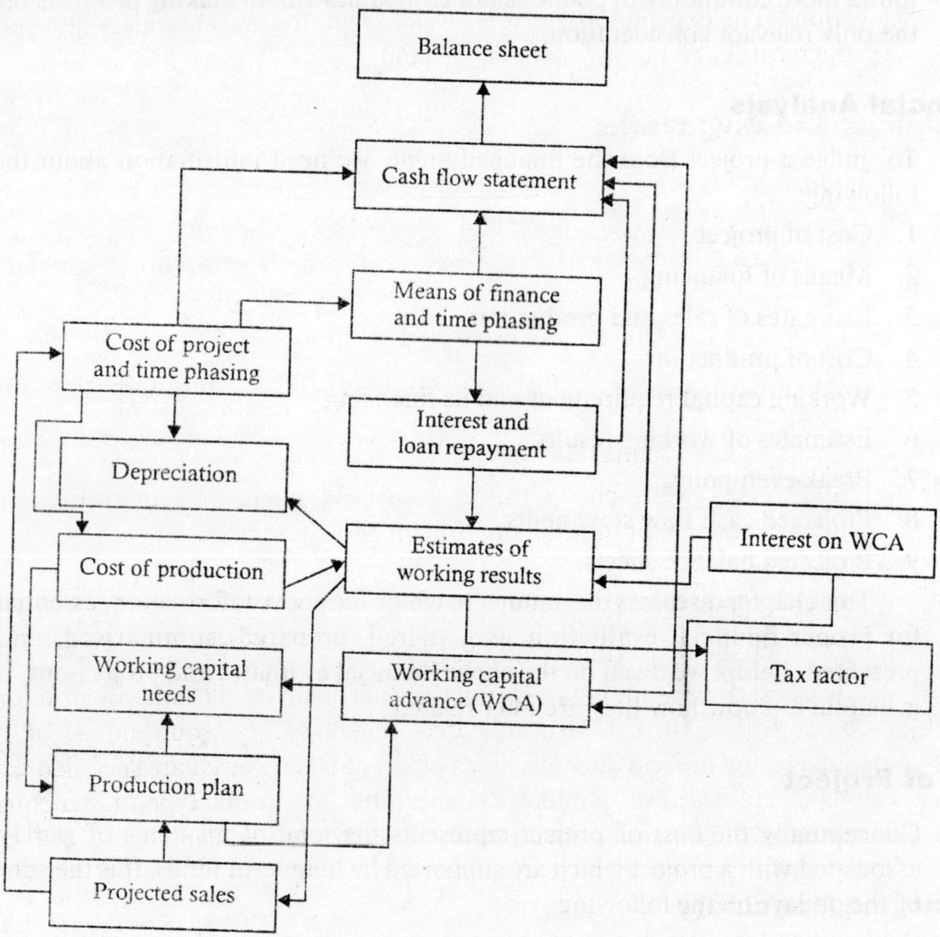

Fig. 2.2

Land and Site Development

The cost of land and site development is the sum of the following:

- Basic cost of land including conveyance and other allied charges.
- Premium payable on lease hold and conveyance charges.
- Cost of levelling and development.
- Cost of laying approach roads and internal roads.
- Cost of gates.
- Cost of tubewells.

The cost of land varies considerably from one location to another. While it is very high in urban and even semi-urban locations, it is reliability low in rural locations. The expenditure on site development, too, varies widely depending on the location and topography of the land.

Buildings and Civil Works

Buildings and civil works cover the following:

- Buildings for the main plant and equipments.
- Buildings for auxiliary services like steam supply, workshops, laboratory, water supply etc.
- Godowns, warehouses and open yard facilities.
- Non-factory buildings like canteen, guest house, time office, excutive house, etc.
- Quarters for essential staff.
- Silos, tanks, wells, chests, basins, cisterns, hoppers, bins and other structures necessary for installation of plant and equipments.
- Garages.
- Sewers, drainage, etc.
- Other civil engineering works.

The cost of the buildings and civil works depends on the kinds of structures required which, in turn, are indicated largely by the requirements of the manufacturing process once the kinds of structures required are specified, cost estimates are based on plinth areas and rates for various type of structures. These rates, of course, vary with the location to some extent.

Plant and Machinery

The cost of plant and machinery, typically the most significant component of project cost, consists of the following:

- Cost of imported machinery: This is the sun of (*i*) FOB (Free on board) value, (*ii*) shipping freight, and insurance cost, (*iii*) import duty and (*iv*) clearing loading, unloading, and transportation charges.
- Cost of indigenous machinery: This consists of (*i*) FOR (Free on rail) cost, (*ii*) sales tax, octroi, and other taxes, if any, and (*iii*) railway freight and transportation charges to site.
- Cost of stores and spares.
- Foundation and installation charges.

The cost of plant and machinery is based on the latest available quotation adjusted for possible escalation. Generally, the provision for escalation is equal to the following product:

(Latest rate of annual inflation applicable to the plant and machinery) × (Length of the delivery period).

Technical Know-how and Engineering Fees

Often it is necessary to engage technical consultants or collaborators from India and/or abroad for advice and help in various technical matters like preparation of project report, choice of technology, selection of plant and machinery, detailed engineering, and so on. While the amount payable for obtaining technical know-how and engineering services for setting up the project is a component of project cost, the royalty payable annually, which is typically a percentage of sales, is an operating expense taken into account in the preparation of the projected profitability statements.

Expenses on Foreign Technicians and Training of Indian Technicians Abroad

Services of foreign technicians may be required in India for setting up the project and supervising the trial runs. Expenses on their travel, boarding, and lodging along with their salaries and allowances must be shown here. Likewise, expenses on Indian technicians who require training abroad must also be included here.

Miscellaneous Fixed Assets

Fixed assets and machinery which are not part of the direct manufacturing process may be referred to as miscellaneous fixed assets. They include items like furnitures, office machinery and equipments, tools, vehicles, railway siding, diesel generating sets, transformers, boilers, piping systems, laboratory, equipments, workshop equipments, effluent treatment plants, fire fighting equipments and so on. Expenses incurred for procurement or use of patents, licenses, trade marks, copyrights, etc., and deposits made with the electricity board may also be included here.

Preliminary and Capital Issue Expenses

Expenses incurred for identifying the project, conducting the market survey, preparing the feasibility report, drafting the memorandum and articles of association, and in corporating the company are referred to as preliminary expenses.

Expenses born in connection with the raising of capital from the public are referred to as capital issue expenses. The major components of capital issue expenses are: under writing, commission, brokerage, fees to managers and registrars, printing and posting expenses, advertising and publicity expenses, listing fees, and stamp duties.

Pre-operative Expenses

Expenses of the following type incurred till the commencement of commercial production are referred to as pre-operative expenses:

1. Establishment expenses.
2. Rent, rates and taxes.
3. Travelling expenses.
4. Interest and commitment charges on borrowings.
5. Insurance charges.
6. Mortgage expenses.
7. Interest on deferred payments.
8. Start-up expenses, and
9. Miscellaneous expenses.

Pre-operative expenses are directly related to project implementation schedule. So, delay in project implementation, which are fairly common, tend to push up these expenses. Appreciative of this, financial institutions allow for some delay (20 to 25%) in the project implementation schedule and accordingly permit a cushion in the estimate for pre-operative expenses. Pre-operative expenses incurred up to the point of time the plant a machinery are set up may be capitalised by apportioning then to fixed assets on some acceptable basis. Pre-operative expenses incurred from the point of time and plant and machinery are set up are treated as revenue expenditure. The firm may, however, treat them as deferred revenue expenditure and write them off over a period of time.

Provision for Contingencies

A provision for contingencies is made to provide for certain unforeseen expenses and price increases over and above the normal inflation rate which is already incorporated in cost estimates.

To estimate the provision for contingencies the following procedure may be followed:

1. Divide the project cost items into two categories *viz*, 'Firm' cost item and non-cost item. (Firm cost items are those which have already been acquired or for which definite arrangements have been made).
2. Set the provision for contingencies at 5 to 10% of the estimated cost of non-firm cost items. Alternatively, make a provision of 10% for all items (including the margin money for working capital) if the implementation period is one year or less. For every additional one year, make an additional provision of 5 per cent.

Margin Money for Working Capital

The principal support for working capital is provided by commercial banks and trade creditors. However, a certain part of working capital requirement has to come from long-term sources of finance, referred to as the 'margin money for working capital', this is an important element of the project cost.

The margin money for working capital is sometimes utilised for meeting over-runs in capital cost. This leads to a working capital problem (and sometimes a crisis) when the project is commissioned. To mitigate this problem, financial institutions stipulate that a portion of the loan amount, equal to the margin money for working capital, be blocked initially so that it can be released when the project is completed.

Initial Cash Losses

Most of the projects incur cash losses in the initial years. Get, promoters typically do not disclose the initial cash losses because they want the project to appear attractive to the financial institutions and the investing public. Failure to make provision for such cash losses in the project cost generally affects the liquidity position and impairs the operations. Hence, prudence calls for making a provision, overt or covert, for the estimated initial cash losses.

Means of Finance

To meet the cost of project the following means of finance are available:

1. Share capital.
2. Term loans.
3. Debenture capital.
4. Deferred credit.
5. Incentive sources.
6. Miscellaneous sources.

Share capital. There are two types of share capitals—equity capital and preference capital.

Equity capital represents the contribution made by the owners of the business, the equity shareholders, who enjoy the rewards and bear the risks of ownership. Equity capital being risk capital carries no fixed rate of dividend.

Preference capital represents the contribution made by preference shareholders and the dividend paid on it is generally fixed.

Term loans. Provided by financial institutions and commercial bank, term loan represent secured borrowings which are a very important source (and often the major source) for financing new projects as well as expansion, modernisation, and renovation schemes of existing firms. There are two broad types of term loans available in India. Rupee term loans and foreign currency term loan. While the former are given for financing land, building, civil works, indigenous plant and machinery, and so on, the latter are provided for meeting the foreign currency expenditures towards import of equipments and technical know-how.

Debenture capital akin to promissory notes, debentures are instruments for raising debt capital. There are two broad types of debentures: Non-convertible debentures and convertible debentures. Non-convertible debentures are straight debt instruments. Typically the carry a fixed rate of interest and have a maturity period of 5 to 9 years. Convertible debentures, as the name implies are debentures which are convertible, wholly or partly into equity shares. The conversion period and price are announced in advance.

Deferred credit. Many a time the suppliers of plant and machinery offer a deferred credit facility under which payment for the purchase of plant and machinery can be made over a period of time.

Incentive sources. The government and its agencies may provide financial support as incentive to certain types of promoters or for setting up industrial units in certain locations.

The incentive may take the form of seed capital assistance (provided at a nominal rate of interest to enable promoter to meet his contribution to the project), or capital subsidy (to attract industries to certain locations) or tax deferment or exemption (particularly from sales tax) for a certain period.

Miscellaneous sources. A small portion of project finance may come from miscellaneous sources like unsecured loans, public deposits, and loaning and hire purchase finance.

Unsecured loans are typically provided by the promoters to bridge the gap between the promoters contribution (as require by the financial institutions) and the equity capital the promoters can subscribe. Public deposits represent unsecured borrowings from the public at large. Leasing and hire purchase finance represent a form of borrowing different from the conventional term loans and debenture capital.

Planning the Means of Finance

We described the various means of finance that can be tapped for a project. How should you go about determining the specific means of finance for a given project? The guidelines and consideration that should be borne in mind for this purpose are as follows:

- Norms of regulatory bodies and financial institutions.
- Key business considerations.

Norms of regulatory bodies and financial institutions

In many countries, including India, the proposed means of financing for a project must be either approved by a regulatory agency or conform to certain norms laid down by the government or financial institutions in this regard. The primary purpose of such regulation is to import prudence to project financing decisions and provide a measure of protection to investors. In addition, the norms of financial institutions, which often provide substantial assistance to projects significantly shape and circumscribe project financing decisions.

Key Business Considerations

The key business considerations which are relative for the project financing decision are: cost, risk, control and flexibility. Sales and production are closely inter-related. Hence, they may be estimated together for this purpose.

Estimates of Production and Sales

(Details may be furnished separately for each product and until the plant reaches maximum capacity utilisation)

	Product				Product			
	1st year	2nd year	3rd year	4th year	1st year	2nd year	3rd year	4th year
1. Installed capacity (quantity per day per annum)								
2. Number of working days								
3. Number of shifts								
4. Estimated production per day (quantity)								
5. Estimated annual production (quantity)								
6. Estimated output as % of plant capacity								
7. Sales (quantity) (after adjusting stocks)								
8. Value of sales (in '000 of Rs.)								

Note: Production in the initial period should be assumed at a reasonable level of utilisation of capacity increasing gradually to attain full capacity in subsequent years.

Cost of Production

Given the estimated production, the cost of production may be worked out. The major components of cost of production are:

- Material cost
- Utilities cost
- Labour cost
- Factory overhead cost.

Materials Cost

The most important elements of cost, the material cost comprises the cost of raw materials, chemical components, and consumable stores required for production. It is a function of the quantities in which these materials are required and the prices payable for them.

While estimating the material cost, the following points should be borne in mind:

1. The requirements of various material input per unit of output may be established on the basis of one or more of the following:
   (*a*) Theoretical consumption norms
   (*b*) Experience of the industry
   (*c*) Performance guarantees and
   (*d*) Specification of machinery supplies.
2. The total requirements of various material inputs can be obtained by multiplying the requirements per unit of output with the excepted output during the year.
3. The prices of material inputs are defined in CIF (cost, insurance, and freight) terms.
4. The present cost of various material inputs is considered. In other words, the factor of inflation is ignored. It may be recalled that the factor of inflation is ignored in estimating the sales revenues too.
5. If seasonal fluctuations in prices are regular, the same must be considered in estimating the cost of material inputs.

Utilities Cost

Utilities consists of power, water and fuel. The requirement of power, water and fuel may be determined on the basis of the norms specified by the collaborators, consultants, etc. or the consumption standards in the industry, whichever is higher.

The cost of power shown here would include only the cost of brought out power, and it may be estimated on the basis of power tariff structure of the concerned electricity boards.

The cost of captive power would naturally be reflected in the cost of fuel, etc. The cost payable to local authorities and charges payable to some other firms for water and/or steam supply may be shown separately. Where the entire water requirement is met out of own wells, water charges need not be shown separately. The cost of fuel (furnance oil, coal, fire wood, bagasse, etc.) often an important item, is somewhat more difficult to estimate.

Labour Cost

Labour cost is the cost of all manpower employed in the factory. Labour cost naturally is a function of the number of employees and the rate or remuneration. The requirement of workers depends on the number of operators/helpers.

Required for operating various machines and manning various services. The number of supervisory personnel and administrative staff may be calculated on the basis of the general norms prevailing in the industry.

In estimating remuneration rates the prevailing rates in the industry/are should be taken in account. The remuneration should include, besides basic pay, dearness allowance, house rent allowance, conveyance allowance, medical reimbursement, leave travel concession, provident fund contribution, gratuity contribution, and bonus payments. In addition, account should be taken of vacations, overtime work, night-work, work on holidays, etc.

Sometimes labour cost is estimated by adding a certain percentage on a global basis to the basic pay. It is, however, advisable to make a detailed analysis, at heart in the beginning.

Labour cost estimates may be raised at the rate of 5 percent per annum to allow for annual increment etc.

Labour cost may be calculated for the year in which the maximum capacity utilisation is first achieved.

For the earlier year, when the capacity utilisation tends to be low, somewhat lower labour costs, but not proportionately lower in relation to capacity, may be assumed.

Factory Overheads Cost

The expenses on repairs and maintenance, rent, taxes, insurance on factory assets, and so on are collectively referred to as factory overheads, repairs and maintenance expenses depends on the state of the machinery. This expenses tends to be lower in the initial years and higher in the later years. Rent, taxes, insurance, etc., may be calculated at the existing rates. A provision should be made for meeting miscellaneous factory expenses, in addition a contingency margin may be provided on the items of factory overheads.

Working Capital Requirement and its Financing

In estimating the working capital requirement and planning for its financing, the following points have to be borne in mind.

1. The working capital requirements consists of the following:
   (*i*) Raw materials and components (indigenous as well as imported).
   (*ii*) Stocks of goods-in-process (also referred to as work-in-process).
   (*iii*) Stocks of finished goods.

(iv) Debtors, and

(v) Operating expenses.

2. The principal sources of working capital finance are:

   (i) Working capital advances provided by commercial banks.

   (ii) Trade credit.

   (iii) Accruals and provisions, and

   (iv) Long term sources of financing.

3. There are limits of obtaining working capital advances from commercial banks. They are in two forms:

   (i) The aggregate permissible bank finance is specified as per the norms of lending, prescribed by the Tandon Committee.

   (ii) Against each current asset a certain amount of margin money has to be provided by the firm.

4. The Tandon Committee has suggested three methods for determining the maximum permissible amount of basic finance for working capital. The method that is generally employed now is the second method. According to this method, the maximum permissible basic finance is calculated as follows:

   Current assets as per the norms laid down by the Tandon Committee non-basic current liabilities like trade credit and provisions.

   The implication of this norm is that at least 25 per cent of current assets must be supported by long term sources of finance.

5. The margin requirement varies with the type of current asset. While there is no fixed formula for determining the margin amount, the ranges within which margin requirements for various current assets lie are as follows:

Table 2.2

Current Assets	Margin
Raw materials	10–25 per cent
Work-in-process	20–40 per cent
Finished goods	30–50 per cent
Debtors	30–50 per cent.

This committee was set up by the Reserve Bank of India to suggest, inter alia, guidelines to commercial banks for working capital financing.

Table 2.3

Items	No. of Month Require-ment	Bank Margin Avail-ability	Amount	Amount of Bank Finance	Margin Money Required	Amount
1	2	3	4	5	6	7
Other items of working capital if any (exercise duty payable on one month's sale, deposits for utilities, etc.)						
Less trade credits available on raw material and consumerables.						
Net working capital.						

Note: 1. The information may be provided for the initial years until the unit reaches maximum capacity utilisation.

2. The amounts given here shall agree with the figures in the profitability statements as well as the cash flow statements.

- Similar information has to be furnished for the II, III and IV years.
- Advance payments, if any, from customers again orders should be deducted from the networking capital and margin money.

Profitability Projections (or Estimates of Working Results)

Given the estimates of sales revenues and cost of production, the next step is to prepare the profitability projections or estimates of working results (as they are referred to by term-lending financial institutions in India) the estimates of working results may be prepared along the following lines:

A. Cost of production.
B. Total administrative expenses.
C. Total sales expenses.
D. Royalty and know-how-payable.
E. Total cost of production (A + B + C + D).
F. Expected sales.
G. Gross profit before interest.

H. Total financial expenses.

I. Depreciation.

J. Operating profit (G – H – L)

K. Other income.

L. Preliminary expenses written off.

M. Profit/loss before taxation (J + K – L)

N. Provision for taxation.

O. Profit after tax (M – N)

Less dividend on

– preference capital

– Equity capital.

P. Retained profit.

Q. Net cash accrual (P + I + L)

Exhibit 6.4 shows form XII of the application form used by all India financial institutions for preparing the estimates of working results.

Cost of production. This represents the cost of materials, labour, utilities and factory overheads as calculated earlier.

Total administrative expenses. This consists of (*i*) administrative salaries, (*ii*) remuneration to directors, (*iii*) professional fees, (*iv*) light, postage telegrams, and telephones, and office supplies (stationery, printing, etc.) (*v*) insurance and taxes on office property, and (*vi*) miscellaneous items.

Total sales expenses. The expenses included under this head are:

(*i*) Commission payable to dealers.

(*ii*) Packing and forwarding charges.

(*iii*) Salary of sales staff (which may be increased at 5 per cent per annum).

(*iv*) Sales promotion and advertising expenses and

(*v*) Other miscellaneous expenses.

The selling expenses depend mainly on the nature of industry and the kind of competitive conditions that prevail typically, selling expenses vary between 5 and 10 per cent of sales.

The experience of similar firms in the industry may be used as a basic guide-line.

Royalty and know-how-payable. Royalty and know-how payable annually may be shown here. The royalty rate is usually 2–5 per cent of sales. Further, royalty is payable often for a limited number of years, say 5 to 10 years.

Total cost of production. This is simply is sum of cost of production, total administrative expenses, total sales expenses, and royalty and know-how payable.

Expected sales. The figure of expected sales are drawn from the estimates of sales and production prepared earlier in the financial analysis and projection exercise.

Gross profit before interest. This represents the difference between expected sales and total cost of production.

Total financial expenses. Financial expenses consist of interest on term loans, interest on bank borrowings, commitment charges on term loan, and commission for bank guarantees. The principal financial expenses, of course, are interest on term loans and interest on bank borrowings.

In estimating the interest on term loans, two points should be borne in mind:

1. Interest on term loans is based on the present rate of interest charged by the term lending financial institutions and commercial banks.
2. Interest amount would decrease according to the repayment schedule of term loan.

Exhibit: Estimates of Working Results.

(This statement should be prepared for ten years)

(in thousands of rupees)

	19	19	19	19	19	Year ending (ten years)
A. Cost of production (as per statement in form XI)						
■ Administrative expenses						
■ Administrative salaries						
■ Remuneration to directors						
■ Professional fees						
■ Light, postage, telegrams and telephones, office						
■ Supplies (stationery, printing, etc.)						
■ Insurance and taxes on office property						
■ Miscellaneous.						
B. Total administrative expenses						
C. Total sales expenses						
D. Royalty and know-how-payable						
E. Total cost of production (A + B + C + D)						
F. Expected sales (as per statement in form XII-A)						

G. Gross profit before interest (F – L)
- Financial expenses
- Interest on term loans
- Interest on borrowings for working capital
- Guarantee commission

H. Total financial expenses

I. Depreciation

J. Operating profit (G – H – I)

K. Other income, if any (give details)

L. Preliminary expenses written off

M. Profit/loss before taxation (J + K + L)

N. Provision for taxation

O. Profit after tax (M – N)
- Less dividend on
- Preference capital
- Equity capital (with rate)

P. Retained profit
- Add depreciation preliminary expenses written off

Q. Net cash accruals.

Note: Detailed working shall be provided for calculation of depreciation (straight line and income tax method), interest, taxation, etc.

The interest on bank borrowings may be estimated as follows:

(*i*) Determine the total requirement of the working capital,

(*ii*) Find out the quantum of bank borrowing that would be available against the total working capital requirement, and

(*iii*) Calculate the interest charges on the basis of the prevailing interest rates.

Depreciation. This is an important item, particularly for capital intensive projects. In figuring out the depreciation charge, the following points should be borne in mind:

1. Contigency margin and pre-operative expenses provided in estimating the cost of project should be added to fixed assests proportionately to ascertain the value of fixed assets for determining the depreciation charge.

2. Preliminary expenses in excess of 2.5 per cent of the project cost (excluding working capital margin) should be added to fixed assets proportionately to ascertain the value of fixed assets for determining the depreciation charge.
3. The income tax act specifies that the written down value method should be used for tax purpose. It further specifies the rate of depreciation applicable to different kinds of assets.
4. For company law (financial reporting) purposes, the method of depreciation may be either the written down value (WDV) method or the straight line (SK) method. From 1988 onwards, the depreciation rates under the companies act have been delinked from those under the income tax act. The key depreciation rates under the companies act are as follows.

Table 2.4

(In percentage terms)

	Single Shift		**Double Shift**		**Triple Shift**	
	WDV	**SL**	**WDV**	**SL**	**WDV**	**SL**
1. Building (other than factory buildings)	5.00	1.63				
2. Factory buildings	10.00	3.34				
3. Plant and machinery (general rate)	15.00	5.15	22.50	8.09	30.00	11.31

Other income. This represents income arising from transactions not part of the normal operations of the firm. Examples of such transactions are: Sale of machinery, disposal of scrap, etc. Except disposal of scrap, which can be reasonably anticipated and estimated, the effects of other non-operating transitions can hardly be estimated. Of course, when non-operating transactions result in a deficit, other income would be negative-put differently, there will be a non-operating loss.

Write-off of preliminary expenses. Preliminary expenses up to 2.5 per cent of the cost of project or capital employed, which ever is higher, can be amortised in ten equal annual instalments.

Profit/Loss before taxation. This is equal to: operating profit + other income – write-off of preliminary expenses.

Provision for taxation. To figure out the tax burden, a sound understanding of the Income Tax Act—a complicated legislation and relevant case laws is required. While calculating the taxable income, a variety of incentives and concessions have to be taken into account. Once the taxable income, as per the Income Tax Act is calculated, the tax burden can be figured out fairly easily by applying the appropriate tax rates. Details pertaining to tax calculation are discussed in an appendix at the end of the book.

Profit after taxation. This is simply profit/loss before taxation minus provision for taxation. A part of profit after tax is usually paid out as dividend—dividend on preference capital and dividend on equity capital.

Retained profit. The difference between profit after tax and dividend payment is referred to as retained profit. It is also called plaughed back earnings.

Net cash accrual. The net cash accrual form operations is equal to: retained profit + depreciation + write-off of preliminary expenses + other non-cash charges.

Review Questions

1. What are the factors to be considered in preparing a feasibility report? Explain its importance.
2. Describe Project Planning and its functions.
3. Explain the different phases of Project Planning.
4. What are the goals and objectives of Project Planning?
5. Give a brief note on Financial Arrangements.
6. Explain the methods used to asses the project profitability.
7. With a neat block diagram to explain the steps involved in financial analysis of a project.
8. Explain:
   (*i*) Cost of Project
   (*ii*) Means of Financing
   (*iii*) Estimates of Sales and Production
   (*iv*) Cost of Production
   (*v*) Working Capital Requirement and its Financing
   (*vi*) Estimate of Working Results
   (*vii*) Break-even Point
   (*viii*) Projected Cash Flow Statements
   (*ix*) Projected Balance Sheets.

3 ORGANISING AND STAFFING THE PROJECT TEAM

SKILLS AND ABILITIES REQUIRED FOR PROJECT MANAGER

Projects are often complex and multidimensional. Managing of these projects or programs represents a challenge requiring skills in:

1. **Team building.** Building the program team is one of the prime responsibilities of the project manager. This involves maintaining:
   (*a*) Effective communications.
   (*b*) Sincere interest in professional growth of team members.
   (*c*) Commitment to the project.
2. **Leadership.** It involves the ability to integrate individual demands, requirements and limitations into decisions that benefit the overall project performance.
3. **Conflict resolution.** This includes understanding the determinants of conflicts so as to respond to the conflicts effectively. Dys functional conflicts result in poor program decision making, lengthy delay over issues and disruption of the team's effort. However, conflicts can be beneficial when 't' produces involvement and new information thus enhancing the competitive spirit. Thus, the manager needs a "Sixth Sense" to determine when a conflict is desirable, what kind of conflict will be useful and how much conflict is optional for a given situation.
4. **Technical skill.** It is essential for the project manager to understand the technology, markets and the business environment to participate effectively in search for integrated solution and technological innovations. Technical expertise is necessary to evaluate technical concepts and solutions to

communicate effectively in technical terms with the project teams and to assess the risks and make trade off between cost, schedule and technical issues.

5. **Planning skill.** It requires the ability to negotiate the necessary resources and commitements from key personnel in the various supporting organisations with little or no formal authority. The planning must be done such that at any point of time in the project, the plan remains viable.
6. **Organisational skill.** The project manager must be a social architect, that is, he must understand how the organisation work, and how to work with the organisation. Organisational skills are particularly important during the project formation when the project manager establishes the project organisation by integrating people from different discipline into an effective work team.
7. **Entrepreneurial skill.** The project manager needs a general management perspective. For example, economic consideration are one component area that normally affects the organisation's financial performance. Thus the project manager must understand the importance of reducing costs in the project.
8. **Administrative skill.** The project manager must be experienced in planning, staffing, budgeting, scheduling and other control techniques.
9. **Management support building skill.** The project manager is surrounded by a variety of organisations that either support him or control his activities. An understanding of these interfaces is important to a project managers as it enhances their ability to build favourable relationship with senior managements.
10. **Resource allocation skill.** Once the project begins, different type of resources are needed at different stages. The project manager must understand this need for varying resources and be capable of allocating resources as and when wanted.

    Thus, the project manager must be both socially and technically aware to understand how the organisation functions and how these functions will affect the project organisation of the particular job to be done. The project manager must also understand the culture and value system of the organisation he is working with.

Requirements of a Successful Project Manager (PM)

The role that the PM will plays is determined by the following:

- The nature of the project—complexity, scale, position in hierarchy of projects;

- The nature of the organisation in which it is being carried out; *e.g.*, sector, activities, organisational structure.
- The personality of the PM.
- The constraints under which they are working.

Trying to specify a generic list of characteristics for a successful project manager would be impossible. The list of characteristics that is presented below should be treated therefore as scales on which:

Initially projects should be rated to determine the requirements of the person for the job.

Individuals can be rated to determine their suitability for the job in hand.

The characteristics of individuals may be classified as either attitudes or skills. **Attitudes** are those characteristics which are determined by the way a person thinks about a particular issue and which are reflected in the intent of that person. The **skills** are those characteristics which a person has been trained in or has learned, and determine tne tools available to that person regarding 'how' to handle a particular situation.

Attitudes which are desirable for project manager are:

- a desire not justify to satisfy but to delight customers and stakeholders alike;
- accepting of both challenges and responsibility;
- being focused on action, rather than procrastination—getting the job done rather than avoiding critical or difficult issues;
- a desire to make the best use of all resources—minimise waste in all activities;
- does not lose sight of the light at the end of the tunnel—is goal focused;
- has personal integrity—people find it very difficult to respect and take the authority of a person who has low integrity;
- is flexible about the route that must be taken to achieve the stated end-goals ;
- has personals goals that are consistent with those of the project organisation the project team-perceives that the PM and the organisation are going the same way.

Skills which are desirable include:

- ability to determine the real/needs desires of the customer. This is done through 'getting close' to the customer *via* visits and both formal and informal discussions, and asking the relevant questions. Analytical skills to turn data into information and break down the project into comprehensible component parts;

- technical skills—the PM need not be a technical specialist, but must at least be capable of corresponding the work that is being carried out, and 'Speaking the Language' of the people involved;
- team skills—many battles have been won against poor odds by the ability of individuals to motivate and enthuse a team;
- ability to delegate effectively—not trying to do everything personally;
- ability to manage your own time—you cannot expect to manage other people unless you can show that you can manage yourself;
- the balancing of stakeholder perceptions of project progress (otherwise known as being also to sell ideas);
- negotiating skills—resolve potential conflict situations so that all parties can be said to have 'won' problem-solving/facilitating problem-solving;

The above are all internal factors of the person. Their role and the way that it is managed by the organisation are just as important. This consideration will centre around four factors;

- responsibility—the scope of the control of the PM as they perceive it;
- authority—the organisation's view as to scope of control of the PM;
- accountability—the set of outcomes that are determined to be under the control of the PM ;
- credibility—the organisation must stand behind the PM and back the decisions that are made. Not doing so removes the individuals credibility and hence authority.

The rules for balancing these roles are:

- no responsibility without authority—do not give a person the apparent responsibility for an area without giving them the authority to carry out their responsibilities.
- no authority without accountability—unless someone can be held to account for their decisions or actions, they should not have the authority to make those decisions.

One of the most common causes of stress among project managers is that they have apparent responsibility for an area, but not the authority to make the necessary changes in order for them to meet the requirements placed on them. This is termed 'role ambiguity'. It is common in organisations that are either only partly structured on projects lines, on for whom the carrying out of project management if any formalised way is relatively novel. There are other stresses that be fall project managers—most notably due to the need. Constantly to meet deadlines and the ever-changing nature of the work. **Kerzner** (1987) suggests that a job advertisement for a project manager should read.

Project manager must have some qualities and facilities for planning and development. He is personable, well-educated, literate individual with degree in engineering to work for small firm. Long hours, no fringe benefits, no security, little chance for advancement are among the inducements offered. Job requires wide knowledge and experience in manufacturing materials, construction techniques, economics, management and mathematics. Competence is required in the use of spoken and written english. There must be more conventional jobs and slanderous insults from colleagues.

Job involves frequent extended trips to inaccessible locations throughout the world, manual labour and extreme frustration from lack of data on which to base decisions.

This is still the case in some companies, but many others have begun to recognise the specialism of project management. The career path of project managers can then be seen to be more than just a string of projects and some discernible progression—as for line managers—is defined. This has only occurred in organisations that have recognised the role of project managment in their corporate strategy.

Project Manager's Authority

A project manager must make decisions to guide the actions of other. This, in brief, is his authority. A project manager, in the first instance, must have authority to enable him, in turn, delegate it to others.

The authority required in project management is for dealing with:

(*i*) Project Scope.
(*ii*) Project Goals.
(*iii*) Project Execution Mode.
(*iv*) Project Organisation.
(*v*) Project Purchase.
(*vi*) Selection of Contractors and Consultants.
(*vii*) Project Technical Performance.

(*viii*) Project Schedules and Budgets.
(*ix*) Funds needed and other resources.
(*x*) Project Personnel.
(*xi*) Public/Share holders.
(*xii*) Project Environment.
(*xiii*) Management system and procedures.
(*xiv*) Project performance review.

On some matters, the project manager will have no authority, and on others he may either have absolute authority or share it with corporate and functional managers. For a company executing projects either regularly or for the first time, it would be necessary for the chief executive to issue what may define the project scope, goals, name the project manager and his authority, name the project reviewing authority and request co-operation of all concerned with project execution. This will lead to the development of the project manual that inturn paves the way. For clearly understanding the authority that wrests with the project manager.

Project Organisations and Types

An organisation chart, in most cases, is the simplest and quickest way to demonstrate the project manager's authority details such as, where the "Project Manager" is positioned, to whom he reports, those with whom he communicates and all those who report to him, tells a lot about the project manager authority. The different types of project organisations are:

1. Line and staff organisation.
2. Divisional organisation.
3. Matrix organisation.
4. Task force organisation.

1. **Line and staff organisation.** In this type of project organisation, a person is appointed with primary responsibilities to coordinate the work in various functional departments. He acts essentially in a staff position to facilitate the co-ordination of the line management is functional departments. This type of organisation is weak because of the influence of the project leader and the relation between the people depend to a great extent on the competence of the project leader. Also, the closeness of the project leader with a few sub-ordinates makes operations difficult in this position the project

manager does not make any decision for the project, nor does the provide any staff service to the functional departments who make all the decisions relating to the project.

2. **Divisional organisation.** Under the type of project organisation, a separate department is setup to implement projects head by the project manager. This division has its complement of personnel over whom the project leader has full line authority. This implies the creation of a separate goal, oriented towards the company's objectives. Division project organisation facilities the process of planning and control that brings better integration of efforts and strength and improves the prospect of fulfilling the time and budget targets. The disadvantage of this setup is that, it is inefficient in using the resources of the firm due to unnecessary duplication of specialists in the company.
3. **Matrix organisation.** The line and functional form of organisation is conducive to an efficient use of resources but it is not suitable for effective use of realisation of project objectives. The divisional form may be suitable for effective realisation but it is not conducive to efficient use of resources, whereas the matrix form is useful to achieve the twin objectives of efficient use of resources and effective realisation of project objectives.

   In a matrix organisation, the personnel working on the project have a responsibility to their functional superior as well as to their project manager. This means that the authority is shared between project managers and functional managers. This type of organisation involves greater complexity and creates a situation when authority and responsibility have a better control in achieving the twin.

   Objective of efficient utilisation of resources and effective attainment of project objectives.
4. **Task force organisation.** In this arrangement, the project manager is delegated the full authority to make decisions on project, but he would be required to operate within the functional organisation's policies and procedures. There is clearly no intervention from the various functional departments no dual decision making and no dual reporting relationship for the workforce. The project manager makes all the decision within the policies and procedures laid down for him. A task force is created by drawing personnel from various functional departments and putting them under the project manager. The staff so assigned will continue to receive administrative support from their home departments but they will respond only to the project manager. While they will receive all directions from the

project managers, they will required to follow the home organisation's policies and procedures.

The organisation structure are as shown below:

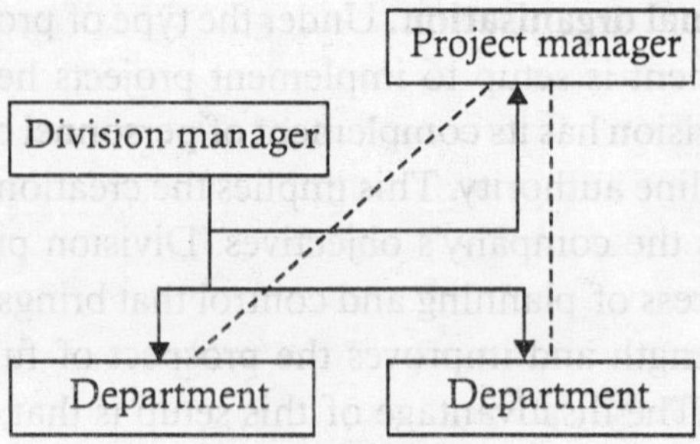

Line-staff

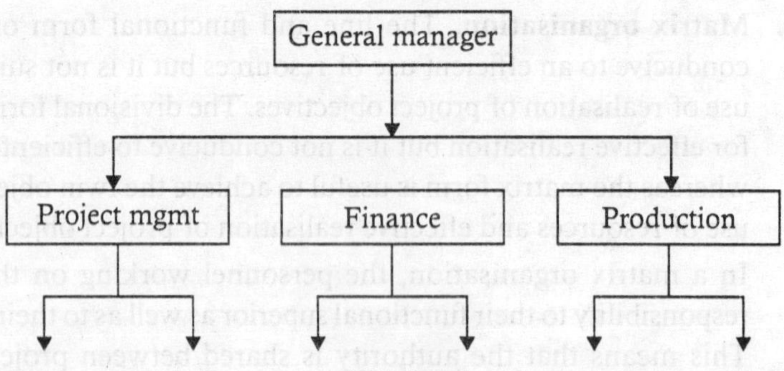

Divisional

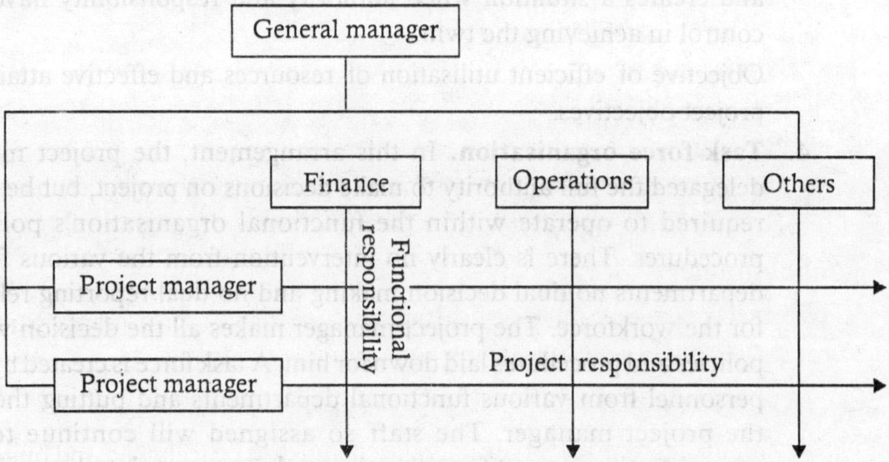

Matrix

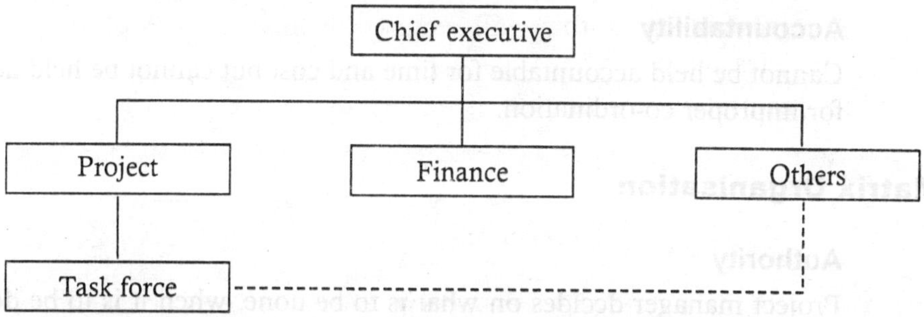

Task force

Accountability in Project Organisation

The structuring of work and delegation of authority for execution of the project would not automatically ensure the achievement of the project. Authority empowers one to make decisions for results in relation to the work at hand. Hence, one who is delegated authority must make decisions and those decisions must produce results. It is meaningless to delegate authority without ensuring that the individuals or agencies will strive for results.

When a person commits himself normally to the achievement of the task whenever he undertakes an assignment or accepts delegation of authority, he can be made responsible by being held accountable for results. If an individual assumes moral responsibility he holds himself accountable to his conscience. But when he accepts responsibility for fear of withdrawal of authority or sanctions of any other form for non-achievement of results, this type of responsibility can be classified as contractual responsibility.

Line and Staff

Authority

The project manager does not make any decision for the project, he collects and communicates information.

Accountability

The project manager cannot be held accountable for any performance parameter of the project. He is accountable only for timely reporting.

Divisional Authority

Project manager decides on the overall schedules to be followed and budgets to be provided but cannot direct staff for adherence to the same.

Accountability

Cannot be held accountable for time and cost but cannot be held accountable for improper co-ordination.

Matrix Organisation

Authority

Project manager decides on what is to be done, when it is to be done and at what cost for the entire project?

Accountability

Can be held accountable for time and cost of the overall project but cannot be held accountable for technical performance.

Task Force

Authority

Project manager decides on what is to be done, when it is to be done, how it is to be done and at what cost for the entire project? He directs the entire project in all aspects of the project but does not recruit, train or involve in corporate administrative matters.

Accountability

Can be held accountable for time and cost and the technical performance of the project.

Tendering and Selection of Contractors

Tender

A tender may be defined as an offer to carry out certain work or supply of certain material or services in accordance with clearly detailed description and conditions.

Contract

In some cases the project manager will have to get work done with resources over which he has no direct authority. It becomes necessary to aquire the required authority in lieu of some considerations. Such an arrangement is termed as a contract. The person chosen to perform the work according to the contractual obligation is termed as a contractor.

A contract presumes that the parties entering into the agreement are competent, normal and clearly aware of all the extremely important for any uncertainty to be resolved at the first instance.

Tendering Procedure

A well laid out procedure for prequalification of contractors.

Procedure for the Tendering and Selection of Contractors

There are four basic steps for the selection of contractors:

1. **Prequalification of contractors.** For prequalification of contractors notifications are issued in the press giving details such as the outline of the project, enquiry issue and tender submission dates, instructions for applying for prequalification and submission dates. Normally, the prequalification document issued on request to a contractor seeks.
   - Information on the organisation.
   - Experience in the intended type of work.
   - Availability of resources like managerial, technical labour and plant.
   - Past turnover and present financial commitments.
   - Credibility with his previous customers.

   The data supplied by the contractors are evaluated for the preparation of a short list. After the short list is completed, the selected contractors are informed and their confirmation obtained as to whether they will submit the tender.
2. **Preparation of tender documents.** A tender document is prepared by the purchaser, in as detailed and clear manner as possible, to define the technical requirement of the work involved, as also the responsibilities which the purchaser and contractor will have to share between themselves. A good tender document usually must include the following:
   - Letter of invitation of tender.
   - Instructions to tenders.
   - Technical specifications.
   - Bill of materials.
   - General conditions of contract.
   - Special conditions of contract.
   - Form of tender.

   This document is then issued to short listed contractors for submission of their tender.

3. **Receipt of tenders.** The tenders may make a request to visit the site normally, the purchaser accompanies the tenderers to the site and provides further information. There may also be a pre-bid conference to clarify various issues of the tenders on the due date, the bids may be opened in front of the tenderers present. The purchaser will announce and record the name of tenderers and prices including prices of alternative tenders. They will also announce and record the names of those tenderers disqualified for late submission.
4. **Evaluation of tenders.** The tenders are evaluated from technical, commercial, contractual and managerial angles. The contractor's clarification and confirmation are sought on many issues that are not very clear in the contract. Normally, separate meetings are held with each contractor to obtain clarification and also to bring all the offers in line with the tender requirement. Actual evaluation process includes checking the acceptability of the offer against technical specifications, management specifications and various commercial contractual terms and conditions. Usually, the lowest bidder who is also technically and managerially accepted is awarded the contract.

Once the contractor has been decided upon an agreement is signed on a stamped paper. The agreement documents include:

1. Original tender papers comprising the conditions of contract dates, specifications, etc.
2. Schedule of rates/prices, including those for engaging workmen, equipment, etc.
3. A list of deviations from original tender stipulations as agreed upon mutually.
4. Other relevant attachments.

Finally a guarantee from sureties may be asked from the contractor as an insurance against uncertainties in dealings with the contractors.

Review Questions

1. What are the skills and abilities, a Project Manager should possess?
2. Enumerate the authorities of Project Manager.
3. Explain different types of Project Organisations with a neat sketch.
4. Differentiate between authority and accountability in different Project Organisation.
5. Define: (*i*) Tender (*ii*) Contract.
6. Briefly explain the procedure for tendering and selection of contractors.

4 PROJECT SCHEDULING

PROJECT IMPLEMENTATION SCHEDULE

Almost the same considerations hold good for project time schedule. At the zero date of the project, we have to set a time target for its completion. This has again got to be realistic because nothing can be achieved by day-dreaming alone.

The problem of data availability apply here as well. For setting of time targets one faces almost the same difficulties as one did in fixing a cost target. At the feasibility stage when the scheduled is first demanded, there is hardly any data to realistically determine the project schedule. And like cost, realistic targets can be arrived at only successively, the final one to be available on completion of the project. Yet despite these apparent difficulties we are for a schedule for feasibility study and at the zero date to set a target and expect that each time it would be the same. But after our discussion on cost estimate, should we expect the schedules to be the same?

There is a qualitative difference between cost estimating and time target setting process. While the need for accuracy in cost estimating is of paramount importance, it is not so with time. Time is not so much dependent on the quantum of work as cost. While the cost of the project can become double with twice the volume of work, the time schedule may not change at all or change only marginally. Moreover, time is a commodity that can be bought with increased cost. The duration of a project can be compressed by putting additional resources on the job. Because of various such reasons accuracy of time estimate has not received as much attention as cost estimates.

Basic of Time Estimate

The cost estimate of a project is made by first making a morphological break down of the project, estimating each component and the summing up the component estimates to arrive at the overall cost estimate. Time estimate, on the other hand, is made by making a work break down of the project, estimating the time schedules for each work, putting them in a proper sequence, *i.e.,* in series or parallel, as per technical or any other logical manner and finally matching their build-up on a time scale with the available resources. The total stretch on the time scale corresponding with the resources sets the targets. Thus, it is not a summing-up process rather it is a manipulating process. Therefore, time is not merely what it takes to do. Work it will also be governed by how much time we want the work to take.

Figure 4.1 illustrates the influence of various factors in setting the time target for a project. The estimated time for the completion of a project is dependent not only on the work content or the sequence, it will also be influenced by resources and constraints. The basic factors *i.e.*, work, constraints and resources are again dependent of the project. The data will get defined/refined successively with the progress of the project. The question that arises, then, is whether we can realistically set the overall project completion target at the beginning of a project. The answer is that we cannot do so if we depend on the manufacturing shop or bottom-up approach for estimating the time duration. In this connection, let us discuss the various approaches possible for estimating the time duration of a project.

1. **Time study approach.** In this approach, the time T for completing a work is

$$T = \frac{Q}{p \times n}$$

where Q = total quantity of work.

p = productivity factor,

i.e., output per man day using normal method of doing work.

n = normal size of crew.

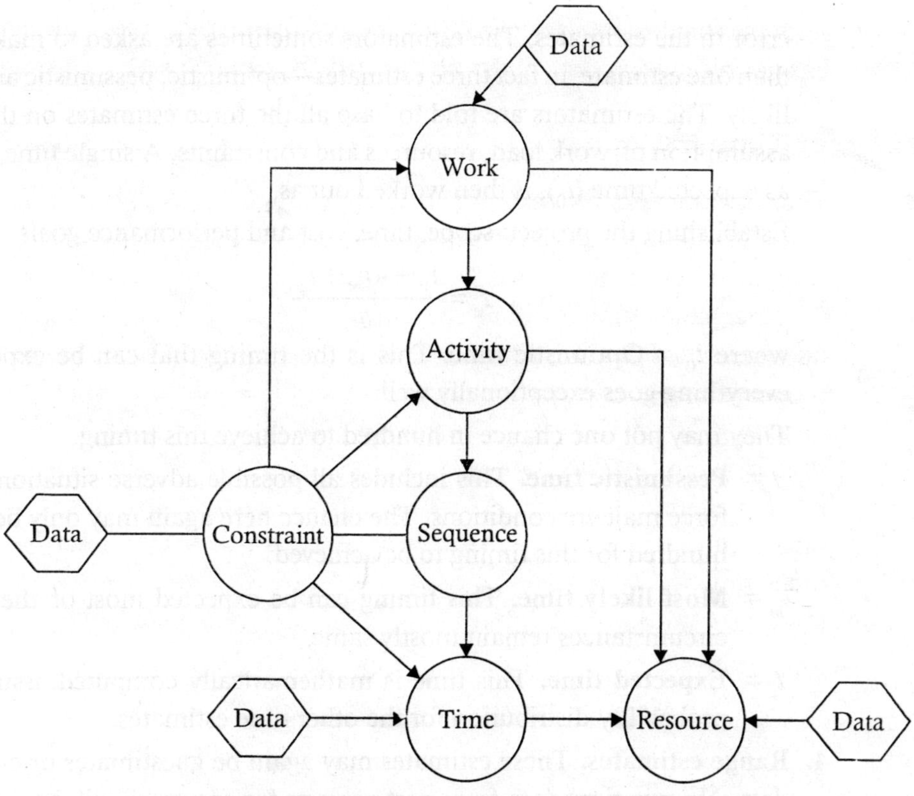

Fig. 4.1 Influence diagram for time estimation of a project.

It can be realized that none of these data are easily available. Factors p and n cannot be ascertained unless they are known from experience of part projects. Also can be estimated correctly only after a project has made substantial progress.

2. **Previous project data.** In this approach data from recently completed projects are used without consideration of Q, p and n. Often these data are used for estimating broad work package durations. Their value are also used for estimating the overall project duration. Past data from projects of similar cost value and complexity are also used. Data from foreign projects may also be used if the same are not indigenously available.
3. **Guestimating approach.** In this approach experienced project personnel are asked to guess project duration more often than not, this is the only approach used in a project for estimating time duration to make up for

error in the estimates. The estimators sometimes are asked to make more than one estimate, in fact three estimates—optimistic, pessimistic and most likely. The estimators are told to base all the three estimates on the same assumption of work load, resources and constraints. A single time, known as expected time (t_0), is then worked out as.

Establishing the project: scope, time, cost and performance goals.

$$t_e = \frac{t_0 + 4t_m + t_p}{b}$$

where t_0 = Optimistic time. This is the timing that can be expected if everything goes exceptionally well.

They may not one chance in hundred to achieve this timing.

t_p= **Pessimistic time.** This includes all possible adverse situation except force majeure conditions. The chance here again may only be one in hundred for this timing to be achieved.

t_m = **Most likely time.** This timing can be expected most of the time if circumstances remain mostly same.

t_c= **Expected time.** This time is mathematically computed assuming a probability distribution for the other time estimates.

4. **Range estimates.** These estimates may again be guestimates or even past data. No two time data from past projects for any work will be the same; they can be better expressed by a range. Vendors quite often quote, deliveries like 6–8 months or 10–12 months indicating a range, though in reality the two figures may be a part of the number game vendors play the short one to given hope to the buyer and the longer one to defend themselves against slippages. Single estimate for any piece of work is arrived at either through computation or negotiation, it is otherwise alien to the whole estimating process.
5. **Time taken *Vs* time required.** What the manufacturing approach of time estimating wishes to establish is the time required to complete a job. But the various methods discussed so far, except the guestimating approach, talk of how much time it has taken to do a job. Actual activity time, true to Parkinson's law, always expands to consume the time allowed. So past data cannot provide a true estimate of the required time. The guestimating process, on the other hand, results in conservative estimates too tight and or too relaxed depending on whether the estimator has any responsibility

for implementation or not. If the estimator has also the responsibility for implementation, he would like to provide all sorts of contingencies to cover up delays which may occur because of change in assumptions regarding scope of work, inefficiency, uncertainties, etc. The reverse happens if the estimator has no responsibility for implementation. The estimator, in that case, tends to take the job too lightly and so provides a very tight and unrealistic estimate.

6. **Estimates from vendors and contractors.** Vendors and contractors are asked to indicate time estimates as they are often asked to quote budgetary cost estimates. These estimates in a competitive situation are supposed to provide a realistic estimate. But this approach provides no better estimates than what has been discussed in the previous paragraph. A vendor may quote for 18 months for a product whose manufacturing cycle-time may be only 6 months. During negotiation, this duration may crash to 6 months or to a slightly higher duration without any extra price. The vendor quotations take together may at best provide a range estimate and are used for the development of a preliminary schedule.
7. **Allocated and committed time.** Except for activities like curing of concrete there are very few activities in a project that would required a fixed duration like an incubation period. The durations could be changed, within a limit, to meet the requirement of the project. So what is necessary in fixing up the project duration is to allocate the time the project could spare and ensure that this allocation is within the limit of comprehensibility of the activity. The allocation must, of course be reasonable, but one need not be too concerned with accuracies.

   In practice, therefore, instead of trying to accurately estimate the duration, a reasonable duration is allocated and commitment obtained from the agency who will be held responsible for implementation. If the duration is not acceptable to any agency, then it may be changed. This may appear to be robbing *x* for paying *y*, as we are asking the implementing agency to absorb the error of the estimator. But as we will see later this is a more pragmatic approach, since many accurate estimates may fail miserably because activities, as has been mentioned before, do not have their own incubation period; they take time because those charged with the responsibility for completing them would not like them to occur otherwise.

What matters is the commitment that could be obtained from implementation agencies, but at the same time it should not be horse trading; that's where the reasonableness of the estimate comes in. The best completed project of the past, either in the same industry or in allied technological area, with a cut in the time estimate to induce challenge could provide the reasonableness one would be looking for in this case. Estimating time duration, therefore reduces people to binding to time durations.

Scheduling for Feasibility Study

In the beginning, there would not be anybody to be bound except the project authorities themselves. So it is up to the project authorities to assume which schedule they want to follow. Ultimately, it is they who will have to ensure its implementation or bear consequences of non-implementation. But since the project authorities stake in any industrial unit may be only 10% of the investment, it is the financial institution who will need to be convinced about the reasonableness of the estimate. It is at this stage that the basis of the estimate will be first questioned, if the project authorities do not choose to do it earlier for themselves.

The data available will be the same as in the case of study estimate for the costs. A typical time schedule for a feasibility study is shown in Fig. 4.2. The schedule almost always is in bar-chart form because the details included are so few and so broad, based that development of a network would not add to any accuracy. Besides this form of presentation suits financial planning and economic evaluation.

It is unfortunate that these schedules are intentionally made extremely tight to show up higher profitability. Despite the fact that these schedules are based on very little information, no provision for error is made. Even the contingency provision which is kept in case of cost estimates to meet uncertainties is not maintained.

Overall Implementation Schedule

To deliver the project is asked to accept the schedule a second look is given to the schedule. This is done first, to confirm the feasibility schedule in the light of enhanced data availability and also to break it down into sufficient details so that commitment from participating agencies on major milestones can be obtained.

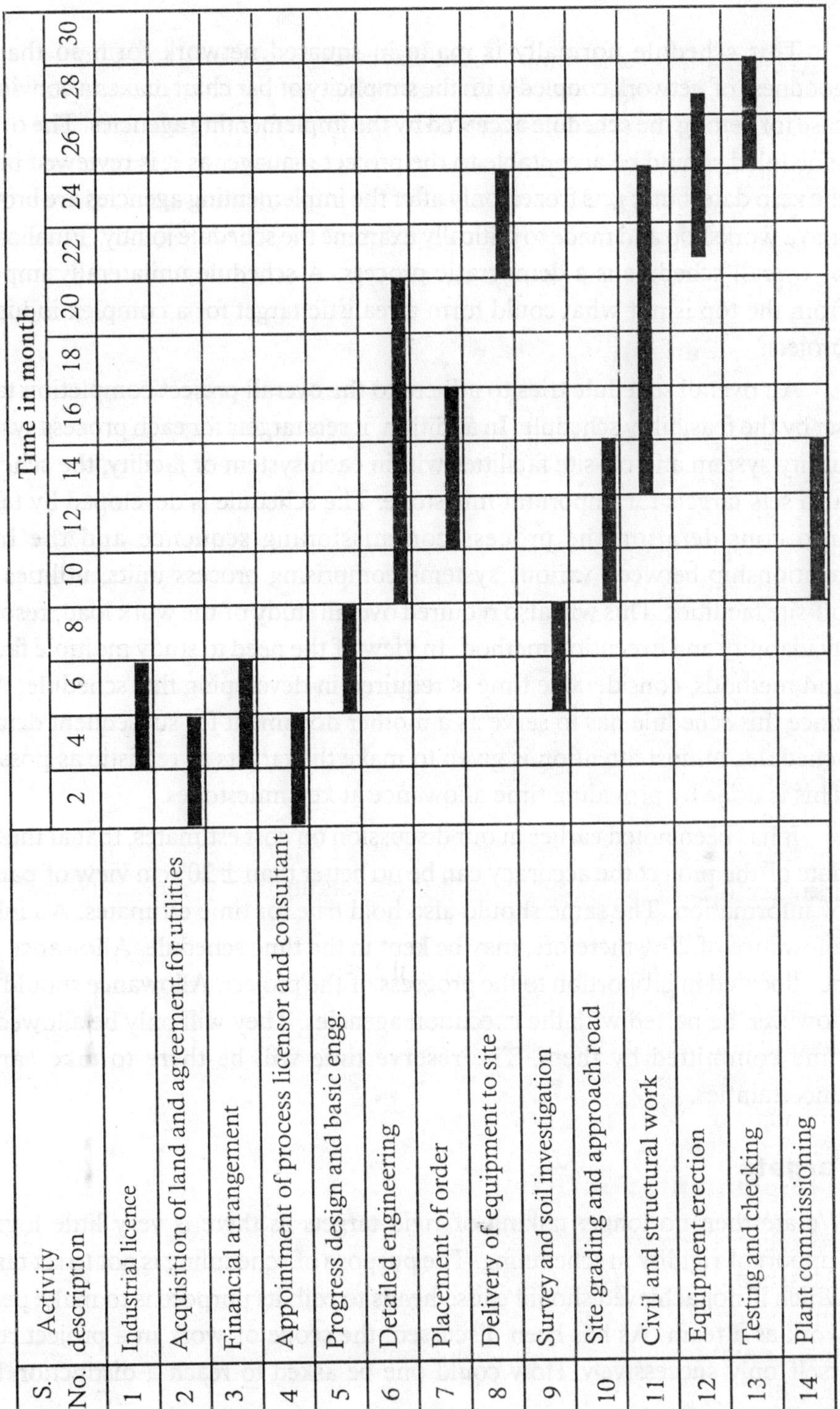

Fig. 4.2. Schedule for feasibility study.

This schedule normally is made in squared network form so that the goodness of network coupled with the simplicity of bar chart makes a convincing case for getting the schedule accepted by the implementing agencies. The overall scheduled should be acceptable to the project manager as it is reviewed before the zero date, but it gets frozen only after the implementing agencies are brought into a workshop and made to critically examine the schedule jointly. Finalization of overall schedule is a democratic process. A schedule unilaterally imposed from the top is not what could form a realistic target for a complex industrial project.

An overall schedule tries to adhere to the overall project completion target set by the feasibility schedule. In addition, it sets targets for each process system, utility system and off-site facilities within each system or facility, the schedule also sets targets for important milestone. The schedule is developed by taking into consideration the process commissioning sequence and the inter-relationship between various systems comprising process units, utilities and off-site facilities. This will also required overall study of the work load, resource availability and execution method. In view of the need to study multiple factors and methods, considerable time is required in developing this schedule. Also, since this schedule has to serve as a mother document for subsequent detailed schedules utmost attention is given to make the targets as realistic as possible. This is done by providing time allowance at key milestones.

It has been noted earlier in our discussion on cost estimates, that at the zero date of the project the accuracy can be no better than ± 20% in view of paucity of information. The same should also hold true for time estimates. An inbuilt allowance of 20% therefore, may be kept in the time schedule. Allowance may be allocated in proportion to the progress of the project. Allowance should not, however, be parted with the execution agencies. They will only be allowed the time committed by them. The reserve time will be there to take care of uncertainties.

Flexi-Targets

We are then no longer talking of rigid targets as there is very little logic in support of rigidity in scheduling. The purpose of scheduling is not to set targets which if not achieved should cause heads to roll; its purpose is to make people work as a team. As has been discussed, the scope of work in a project revels itself only successively. How could one be asked to reach a distinction by a certain time without knowing the path? When a schedule is developed something is known, something half-known and the rest is all unknown. A schedule provides the much needed visibility in this journey through the unknown. It provides direction, makes people see things which otherwise cannot be seen and consequently enables people to communicate and make commitments. It provides visibility in the process of team work and so enables management to co-ordinate and direct the efforts of the project team on the right course. The target is at best a forecast of possible completion time. Since there is a possibility of achieving the target, every effort must be made to achieve it. If on a second reckoning the old target does not appear feasible there is no point in maintaining it. The target to serve as a base line for performance measurement must not be beyond achievement. Time targets, as in the case of cost targets, need to be revised at different stages of project management to serve the purpose for which they are meant. It must undergo one such revision on the zero date to make the start of the project on a realistic note.

TIME MANAGEMENT

Managing projects within time, cost and performance is easier said than done. The project managment environment is extremely turbulent, and is composed of numerous meetings, report writing, conflict resolution, continuous planning and replanning, communications with the customer, and crisis management.

In the project environment there are numerous opportunities for the project manager to get into trouble. The list below identifies the most common time management traps that can defer one from effective project management:

- Incomplete work
- A job poorly done that must be done over
  - Delayed decisions
- Poor communications channels
- Uncontrolled telephone calls
  - Casual visitors
  - Waiting for people
  - Failure to delegate
- Poor retrieval system.

The problems that exist in trying to obtain effective project time management. Fifteen areas were investigated.

- Employees's background age, project time span and value of project
- Energy level per day
- Energy level per week
- Daily/weekly work schedule
  - overtime
  - productivity
- Meetings
- Time robbers
- Time away from desk
- Priorities
- Communications
- Conflict management
- Planning/replanning
- Community service
- Delegation
- Productivity.

An individual can have a large amount of energy but the productive use of this energy may be low. The average employee is usually productive about six hours in an eight hour work-day. The reason for this 70–80 per cent productivity level is the various items "Steal" time needed for the completion of a task.

The following such items were identified by project managers:

- Lack of information in a ready format
- Phone cells
- Day-to-day administration
- Spending more time than anticipated in answering questions
- Lack of sufficient clerical support
- Late appointments
- Impromptu tasks
- Unscheduled appointments or "drop-ins".
- Union grievances

- Having to explain "thinking to superiors".
- Too many levels of review
- Too many people in a small area
- Office casual conversation.

Rules for Time Management

- Conduct a time analysis (time log)
- Classify your activities
- Establish priorities
- Establish opportunity cost on activities
- Train your system (boss, subordinates, peers)
- Practice delegation
- Practice calculated refusals
- Practice management by exception
- Focus on opportunities—not on problems.

Question

- What am I doing that I don't have to be doing at all?
- What am I doing that can be done better by someone else?
- What am I doing that could be done sufficiently well by someone else?
- Am I establishing the right priorities for my activities?

Time Management Recommendations for Project Managers

- Know the weekly and daily energy cycle of your people as well as your own. Be sure to assign or perform work that is compatible with this energy cycle.
- If you have employees who come and go on flex-time schedules, be sure to account for this arrangement in assigning work and understanding their energy cycle.
- Understand the productivity levels of your people, and make sure that the projects performance standards are compatible with the productivity level of your people.
- Do not schedule overtime unnecessarily unless you know that overtime is needed and that efficiency will be maintain it is possible to employees to

"save themselves for overtime" and thereby produce the same work in twelve hours that they would in eight hours.

- Try to monitor your own workload closely and see if there is any work that does not have to be done effectively by someone else. If necessary, refuse to do the unimportant work. Avoid procrastination and try to work on the most difficult task first, start now and look for ways to buy additional time. Be prepared to make quick decisions.
- Do not schedule meetings unless they are cost-effective and necessary. Assist your people in preparation for the meeting. Prepare an agenda and make sure that key personnel are informed well in advance of any major problems to be discussed.
- Conduct the meeting effectively and efficiently. Start the meeting on time, get right to the point, and end the meeting in time. Try to get all attendes to express their views, and avoid prolonged discussions of trivial tasks.
- Decide whether or not it is absolutely necessary for you to attend a given meeting, especially if it requires travel time. If travel time is required, be prepared to work at travel stops.
- Try to minimize the amount of time you spend away from your desk. Be willing to delegate. Plan solid blocks of time for important work. Classify your objectives, and get the point at once. Learn how to say no. Be willing to delegate and employ the management-by-exception concept.
- Time robbers can destroy a good project schedule. Control telephone time and be willing to let your secretary take messages. Get rid of casual visitors. If necessary, find a way to work in seclusion.
- Eastablish proper priorities for yourself, your project, and your people.
- Avoid time-consuming communication processes, avoid memos. If memos or letters are necessary, make them short and summary-type in nature. If you have lengthy reports to read, it is best to take them with you on long trips.
- Train your boss, peers and subordinates and how to work with you. Be willing to assert you rights.
- If conflict resolution is necessary, obtain the necessary information as fast as possible and make a decision. Eastablishing procedures to conflict resolution may be helpful.
- Follow your schedules closely, especially items on the critical path. You may find it necessary to monitor critical items yourself rather than wait for periodic feedback.

Project Management

- Be willing to delegate work to subordinates and peers. Do not try to be a "nice guy" and do it all yourself, let you place yourself in the position of doing work that is normally the responsibility of the functional departments or other project office personnel.
- Project managers typically understand well the role of the project manager at project conception, but seem to forget it during project execution. This loss of understanding which creates time management problems for the project manager is usually caused by the project manager's.
- Waiting for someone else to make a decision that is his own responsibility.
- Neglecting to "keep his door open" and "walk the halls" to find out what's going on.
- Neglecting to use the experts correctly and trying to do it all himself.
- Being concerned about his previous technical discipline or profession rather than the best interest of the company.
- Being too interested in methods rather than in results.
- Trying to do the work himself rather than to delegate it to someone working slower.
- Wasting time in project team meetings discussing one-on-one problems.
- Failing to recognize that his boss is there to help.

Project managers must understand that even though they have the authority, responsibility, and accountability for a project, there are still parent company administrative duties that must be accepted. These items are usually additional work that project manager has not considered .

Project management may not be the best system for managing resources but it is better than anything we have in the past. Effective time management may very well be the most important weapon in the project manager's arsenal for obtaining proper resource control.

Management Pitfalls

The project environment offers numerous opportunities for project managers and team members to get into trouble, above and beyond time management traps. These problem-causing activities are referred to as management pitfalls. Common types of management pitfalls are:

- Lack of planning
- Lack of self-control (knowing oneself)

- Activity traps
- Managing *Vs* doing
- People *Vs* task skills
- Ineffective communications
- Management bottlenecks.

Meetings

Meetings become a way of life in project management and project engineering. Unfortunately, meeting can become un-productive and a total waste of time especially if the project manager:

- spends too much time on trival items
- neglects sending out an agenda
- holds too many team meetings
- holds too few team meetings
- neglects inviting personnel with decision-making authority.

Time Robbers

In the previous discussions, identification was made of the most common time robbers in a project environment.

These included:

- Incomplete work
- Jobs done over
- Delayed decisions
- Poor communications
- Telephone calls
- Causal visits
- Waiting for people
- Failure to delegate
- Poor retrieval system
- Others.

The project mangers were querried as to which of these time robbers commonly occur on their projects, and how much time is usually spent per weak on each time robber. Delayed decisions and poor communication were the most commonly identified items. Most of the industries appear to have two or three predominant time robbers, except for the construction project managers.

Most respondents identified "other" items that detracted from their performance seventy per cent of the major automotive equipment manufacturer respondents identified such items. The following "other" items were listed in this survey:

- Lack of information in a ready format
- Too many people in a small area
- Casual office conversations
- Day-to-day administration
- Spending more time than anticipated answering questions
  - Lack of sufficient clerical support
  - Late appointments
  - Impromptu tasks
- Too many level of review
- Union grievances.

Effective Time Management

There are several techniques that project managers can practice in order to make better use of their time.

1. Delegate.
2. Follow the schedule.
3. Decide fast.
4. Decide who should attend.
5. Learn to say no.
6. Start now.
7. Do the tough part first.
8. Travel light.
9. Work at travel stops.
10. Avoid useless memo.
11. Refuse to do the unimportant.
12. Look ahead.
13. ASK: Is this trip necessary?
14. Control telephone time.
15. Send out the meeting agenda.

16. Shut off in house visits.
17. Overcome procrastination.
18. Manage by exception.

Scheduling

Scheduling may be defined as the assignment of work to the facility with the specifications of time and the sequence in which the work is to be done. The facility may be manpower, machine or both. According to Kimball & Kimball, '*scheduling is the time that is required to perform each operation and also the time required to perform the entire series as routed*'.

The principle aim of scheduling is to plan the sequences of work so that production can be systematically arranged towards the completion of all products by due date.

Scheduling is basically a detailed day-to-day planning of operations. It deals with questions such as:

(*a*) Which work centre will do which job?

(*b*) When should an operation/job be started? When should it end?

(*c*) On which equipment should it be done and by whom?

(*d*) What is the sequence in which job operations need to be handled through a set of machines or work centres?

A production schedule is similar to a engineering class time-table which gives us information on which class is to be held by when and for what duration and how much to be taught and so on.

Objectives of Scheduling

1. Scheduling aims to achieve the required rate of output with a minimum of delay and disruption in processing.
2. To produce enough so as to maintain finished inventories to level predetermined to meet delivery commitments.
3. The aim of scheduling is to have maximum utilization of men, machines and materials by maintaining a free flow of materials along the production line.
4. To prevent unbalanced allocation of time among work centres with a view to eliminate idle capacity.
5. To keep the production cost minimum.

Scheduling Strategies

Scheduling strategies differ widely from organization mainly depending on the product produced, size and type of production, company policies and priorities etc. Some companies have 'no scheduling' at all, while some companies have very sophisticated approaches to scheduling. It is noted here that most scheduling strategies are concerned with job shop production since the problem of scheduling is more encountered in jobbing industries. It is convenient to classify the scheduling strategies into four groups:

1. Detailed scheduling.
2. Cumulative scheduling.
3. Cumulative detailed scheduling.
4. Priority decision rules.

1. **Detailed scheduling of specific jobs.** This means that all the job orders that arrive from customers are immediately scheduled to the last detail as and when they come in. This is quite impractical in a manufacturing environment. Normally schedules are subject to disruptions due to reasons like absenteeism of workers, break down of machinery, order-attractions by customers or any other unexpected events at any work centres. Therefore a detailed scheduling often becomes uncontrollable for the production manager.

   However detailed scheduling is used efficiently in service industries such as airline, hotels, consulting doctors, etc. In such places, detailed scheduling of orders is done well in advance of rendering actual service.

2. **Cumulative scheduling of the total work load.** This means that all customer orders for various jobs are pooled together to form a cumulative work load. This total work load is then matched with the capacity available and then allocated broadly is such a way that immediate periods gets allocated to maximum capacity.

   Cumulative scheduling of the total work load is useful, especially for long range planning of approximate capacity needs. However, this type of scheduling may result in the overloading of current periods and the under loading of future periods. It offers no guidance as which jobs to do when and there is no means of controlling which orders are being worked on.

3. **Cumulative-detailed combination.** Cumulative-detailed approach to scheduling is a feasible and practical approach which combines the 'firm'

and 'flexible' natures of the two earlier strategies namely detailed scheduling and cumulative scheduling respectively. This could actually be in tune with the requirements of the master production schedule which might have both 'firm' and 'flexible' portions of requirements.

Cumulative-work load projections, made for jobs arrived, can be used to plan for approximate capacity needs. Capacities are planned on a broad basis first in terms of total labours and machine-hour requirements per week at key work centres. As changes occur during the weeks prior to manufacturing, the material and capacity requirement are updated, usually with the help of a computer automatically. Capacity may be allocated to specific jobs later, perhaps just a few days before the work is to be performed. Normally one-day is the shortest practical scheduling unit used in job shops. The actual time allowed for the job. Even at each work centre is based on predetermined standard hour requirements for that job. Even the exact hour and minute of the working of each job can be pin-pointed but it would be normally difficult to stick to such razor-sharp schedules.

4. **Priority decision rules.** When there are a set of orders to be executed the question of which to be done first always arises. Different organizations have different criteria for deciding the priority of jobs to be processed. These priority decision rules are scheduling guides that are used jobs to be processed. These priority decision rules are scheduling guides that are used independently or in conduction with any one of the above strategies. Examples of such rules are 'first-come first served', 'last-come first-served' 'random selection', etc.

   When used independently, priority decision rules are like any other scheduling strategy when used in conjuction with the combination of cumulative and detailed scheduling priority rules can be very effective. One such way of using priority rules with other strategers is the critical ratio concept, which is a very widely used rule.

Forward and Backward Scheduling

There are two more approaches to scheduling. Forward scheduling is an approach where the customer are immediately processed as soon as they come in even if their due dates are far away. In other words, forward scheduling starts as soon as end item requirements are known and often results in their completion well before the required due date. For example if a component is required to be delivered in two weeks time and if it needs a total of nine days to process it, the first nine days are used to make it. The component could be delivered immediately after completion or would rest as a finished goods inventory for the remaining 5 days before it is delivered (Figure 4.3).

Backward scheduling. It is an approach where the customer orders are processed as late as possible so that they are finished and delivered exactly on their due date. In other words, in backward scheduling the starting time of the processing of the job is determined by 'setting back' the number of days required for its processing, from the finish date. Consider the same components that requires 9 days of processing and needs to be delivered at the end of two weeks. This component is commenced to be processed from the 6th day onwards so that it is just complete on the 14th day and delivered immediately.

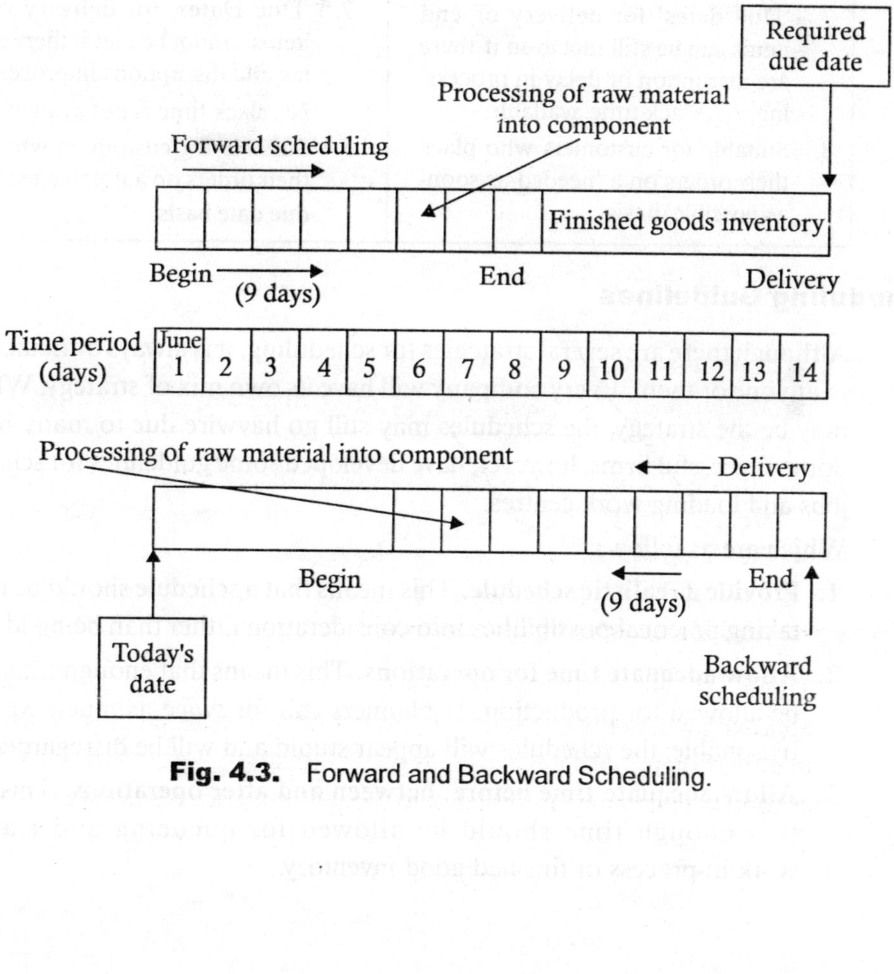

Fig. 4.3. Forward and Backward Scheduling.

Forward *Versus* Backward Scheduling

Forward Scheduling	Backward Scheduling
1. Components are delivered as soon as possible'.	1. Components are delivered only 'when needed'
2. Start and finish times for waiting jobs are determined by assigning them to the earliest available time slots at the work centre.	2. Start and finish times for waiting jobs are determined by assigning them to the latest available time slots at the work centre.
3. Components are expected to be completed before their due date.	3. Components are expected to be completed exactly on their due date.
4. More work-in-process inventory.	4. Less work in process inventory.
5. Higher inventory carrying cost.	5. Less inventory carrying cost.
6. More storage space required.	6. Less storage space required.
7. 'Due dates' for delivery of end items can be still met even if there are disruption or delay in processing, *i.e.*, slack time available.	7. 'Due Dates' for delivery of end items cannot be met if there are delay and disruptions in processing, *i.e.*, slack time is not available.
8. Suitable for customers who place their orders on a 'needed-as-soon-as-possible' basis.	8. Suitable for customers who place their orders on a not needed before due date basis.

Scheduling Guidelines

Although there are several strategies for scheduling, it is always difficult to stick to anyone of them. Every company will have its own mix of strategy. Whatever may be the strategy, the schedules may still go haywire due to many reasons. Some successful firms, however, have developed some guidelines for scheduling jobs and loading work centres.

Which are as follows:

1. **Provide a realistic schedule.** This means that a schedule should be realistic taking practical possibilities into consideration rather than being idealistic.
2. **Allow adequate time for operations.** This means that enough time should be allowed for production. If planners call for twice as much work as is reasonable, the schedules will appear stupid and will be disregarded.
3. **Allow adequate time before, between and after operations.** This means that enough time should be allowed for queueing and transit of work-in-process or finished good inventory.

4. **Don't release all available jobs to the shop.** Releasing all the available jobs as and when they are received overloads the capacity, it also increases manufacturing load times and excess work in process inventory.
5. **Don't schedule all available capacity.** In the shop, good scheduling systems always release capacity at a reasonable rate to the available jobs. Some excess capacity should always be available to handle emergencies and sudden alteration in jobs or to accommodate a totally new but profitable customer order.
6. **Load only selected work centres.** This means that only those work centres which are full fit should be selected for operations. Even if all the work centres are full fit, a few should be free to handle emergencies and maintenance work.
7. **Allow for necessary changes.** Schedules should always be open to necessary changes and alterations in the products which is quite often the case.
8. **Gear-up the entire shop to be responsible for the schedule.** It is not only the workers responsible for the production, but the co-operation of the entire work-force which is necessary to stick to schedules which are planned. So everyone in the industry is responsible to the schedule in one way or the other.

Scheduling Methodology

There are several scheduling methods that are practised in the industry. However the specific method used depends upon the type of industry, organization, product and level of sophistication required.

Scheduling methods can be generally classified as

I. Charts and boards

1. Gantt load chart.
2. Gantt progress chart.
3. Schedule boards.
4. Computers graphics.

II. Priority decision rules

1. Single criterion rules.
2. Combined criteria rules (Johnson's rule).
3. Critical ratio scheduling.
4. Index method of scheduling.
5. Critical path method.

III. Mathematical programming methods

1. Linear programming: simple and transportation methods.
2. Assignment method.
3. Dynamic programming.

I. Charts and Boards

Gantt charts and scheduling boards have been extensively used ever since Henry Gantt developed them in 1917. Gantt charts were manually prepared and used in scheduling in job shops but nowadays computer generates all the charts and boards much more quickly and clearly more importantly computers charts and graphics can be developed always on a up-to-date status, when immensely helps in controlling scheduling operations.

1. **Gantt load chart.** A Gantt load chart shows the amount of cumulative workload that each work centre has in a manufacturing unit. It is nothing but a graph showing workloads of each work centre on a time-scale. In other words, the chart shows the total estimated workload ahead of each work. Centre for all the jobs to be done. However, the chart does not specify which job will be completed at which time, nor does it show the sequence in which the jobs should be processed. While drawing the Gantt load chart, no allowances are made for any delays or disruptions due to break down, absenteeism maintenance at the work centres the chart just shows the cumulative workload for each work centre.

   Consider a facility where there are four work centres and five jobs. These five jobs (A, B, C, D, E) require processings on one or more among the four work centres. The total cumulative load for each work centre is then drawn in the form of a Gannt load chart as shown in Fig. 4.4.

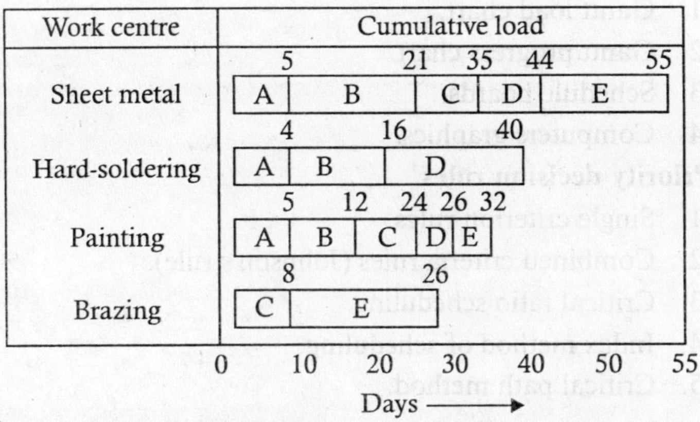

Fig. 4.4. Gantt load chart.

Thus, the chart shows that 55 days of cumulative work lie a head for the sheet metal centre, the paint centres faces a 32 day load, and so on. But it can be observed here that job A cannot be processed simultaneously at more than one work centre. It has to be processed according to the sequence of operation as required by the job and therefore the total time required to do all the jobs by all work centres will be much more than 55 days.

Uses of a Gantt Load Chart

(*i*) Total work load for each work centre is shown graphically which is simple, clear and easy to understand.

(*ii*) It indicates the need for more resources or for reassigning or resources when the load at one work centre becomes too large.

(*iii*) In case some work centres are overloaded, employees from a low-load work centre may be temporarily shifted to high load centres.

Limitations

(*i*) Gantt load chart does not account for delays and disruptions at the work centres due to any reason.

(*ii*) It does not give information regarding the due data requirements of each job.

(*iii*) The chart must be updated periodically to account for new jobs.

2. Gantt schedule and progress chart. A gantt schedule and progress chart basically indicates the scheduled starting and completion dates for each jobs as well as the current status of each jobs as on today. In other words, the chart shows an immediate comparison between the schedule and reality with respect to each job and work centre. The following symbols are used while drawing a gantt schedule and progress chart.

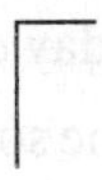 — Indicates date or hour of scheduled starting time of job.

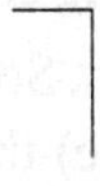 — Indicates data or hour of scheduled completion time of each job.

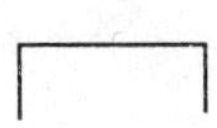 — Indicates scheduled duration of completion of each job.

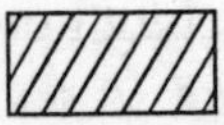 — Length of band indicates the relation of completed work to scheduled work *i.e.*, current status of the job.

 — Indicates a duration of unavoidable delay due to break down, absenteeism, etc.

 — Indicates today's date.

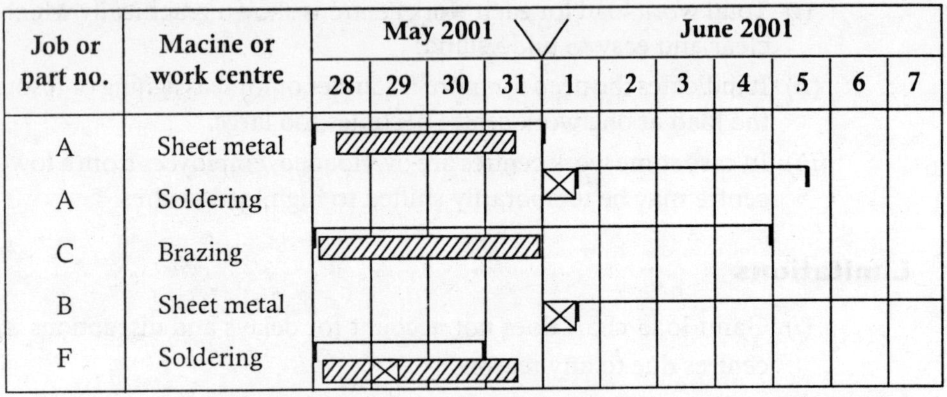

Fig. 4.5. Gantt schedule and progress chart.

The gantt schedule and progress chart shown in Fig. 4.5 can be understood through the following points:

(*i*) Job A is being processed through the sheet metal centre. Its scheduled starting date is 28th May, scheduled completion plate is 31st May and the scheduled duration of work is obviously of days. Today is 1st of June, and as indicated by the length of the band, the job A is not completed with half a day of work still remaining to be done.

(*ii*) After sheet metal work is over, the job A required half a day of change over (indicated by ) for the continuation of processing in the soldering centre. The scheduled duration of work for job A in the soldering centre is A days.

(*iii*) Let us assume that job C starts with the brazing operation. So it is scheduled between 25th May and 4th June. As on today (1st June) the job is up-to-date and not lagging behind as indicated by the band in other words, job C is on schedule.

(*iv*) Sheet metal centre is scheduled to be free only from 1st June onwards because it is occupied with job A. After the completion of job A, sheet metal needs half a day for maintenance. Therefore job B is scheduled for sheet metal from the afternoon of 1st June. It is to be noted here that if job A is delayed, job B will also be delayed.

(*v*) Job F was scheduled between 28th May and 30th May. But there was an unavoidable delays of half a day and therefore the job was completed half a day behind schedule.

Thus a gantt schedule and progress chart is of immense use in scheduling jobs. It help in planning and replanning activities and is invaluable to the production department.

3. **Schedule boards.** It is not enough if only the management observes and operates the gantt charts. Everyone on the shop floor needs to know this information so that they know where they stand with respect to the various jobs. Big schedule boards showing gantt charts can be put up at each department which has a motivating effect on the workers. Numerous mechanical and magnetic charts and boards are available to faciliate easy comprehension of various schedules. Examples of such boards used in the American industry are board master, sched-u-graph, etc.
4. **Computer graphics.** Computer reports and graphics have vastly replaced the mechanical boards and charts computer graphics are also being increasingly used in the Indian industry. Computer packages are of great help because they can keep track of thousands of item so inventory and assist in generating revised schedules very quickly.

II. Priority Decision Rules

Every jobbing industry is always faced with the problem of priority decision with respect to the jobs on hand. In other words, among all the customer orders or waiting jobs, which should be done first? This question can be answered by several priority decision rules that are practised in the industry. A priority decision rule is a systematic procedure for assigning priorities to waiting jobs, thereby determining the sequence in which jobs will be processed. In other words priority decision rules are simplified guidelines which lays down the basis for determining the sequence in which job will be done. It should be noted here that no single rule is best for all situation and different industries use different rules according to situations however some major criteria for reflecting a particular rule would be:

(*i*) Set up costs.

(*ii*) In-process inventory.
(*iii*) Idle time of machines and labour.
(*iv*) Percentage of jobs that are late.
(*v*) Average number of jobs waiting in the queue.
(*vi*) Average time to complete a job.
(*vii*) Standard deviation of time to complete a job.

Priority decision rules can be classified as:

1. Single criterion rules.
2. Combined criteria rules.
3. Critical ratio scheduling.
4. Index method of scheduling.
5. Critical path method.

I. Single Criterion Rule

Some of the simplest but sometimes the most effective priority rules to assign jobs are on the basis of a single criterion. Some of these rule are as follows:

1. **FCFS—First Come, First Served.** This is a priority rule that given top priority to the waiting job that arrives earliest in the production system. In other words, incoming jobs are processed in their order of arrival. FCFS is commonly applied in service industries such as banks, post offices rather than in manufacturing industries.
2. **EDD—Earliest Due Date.** This is a priority rule that gives top priority to the waiting job whose due date is the earliest. In other words, the job is taken up first whose scheduled due date of completion in the earliest EDD actually ignores the order in which jobs arrive and their processing time.
3. **LS—Least Slack.** This is a priority rule that gives top priority to the waiting job whose slack time is least. Slack time is calculated as the difference of the length of time remaining until the job is due and the length of its operation time. For example, if a job requires 8 days of processing and if there are 10 days left for its delivery data, slack time would be (10 – 8) = 2 days.
4. **SPT—Shortest Processing Time.** This is a priority rule that gives top priority to the waiting job whose due date is the earliest. In other words, the job that can be completed in the shortest time at the work centres is processed next. The order of arrival and due dates of jobs are ignored.

5. **LPT—Longest Processing Time.** This is a priority rule which gives top priority to that job which takes the longest time to be processed.
6. **PCO—Preferred Customer Order.** This is a priority rule that gives top priority to orders that come from some favourite customers. The order from preferred customers are done first irrespective of their order of arrival or processing times.
7. **RS—Random Selection.** This is a priority rule that selects jobs randomly. In other words there is no particular basis on which jobs are selected and is a pure chance for any job to be selected first. It is just like a lottery.

Worked Problems

1. There are five jobs which are waiting to be processed at the sheet metal work centre. The jobs have arrived in the alphabetical order. The processing times and due dates are given below.

   (*i*) Determine the sequence of jobs using FCFS and SPT rules.

   (*ii*) Find total completion time, average flow time, average number of jobs in the system each day, and average job lateness for each of the above rule.

Job	A	B	C	D	E
Processing time (in days)	4	17	14	9	11
Due date (in days from now)	6	20	18	12	12

Solution:

(*i*) **Sequence of jobs using FCFS rule.**

Waiting Job (in FCFS sequence)	Processing time (in days)	Flow time (in days)	Due date (in days from now)
A	4	4	6
B	17	21	20
C	14	35	18
D	9	44	12
E	11	55	12
Total	55		

(*a*) Total completion time = 55 days for all jobs.

(*b*) Average flow time $= \dfrac{\text{Sum of flow times for all jobs}}{\text{Number of jobs}}$

$$= \frac{4+21+35+44+55}{5} = 31.8 \text{ days.}$$

The above quantity means that on an average a job is in the system for 31.8 days. In other words, it is the average of the sums of the waiting times and processing times of all jobs.

(*c*) Average number of jobs in the system each day. It is the average number of jobs flowing in the system (waiting or being processed) from the beginning of the sequence through the time when the last job is finished. That is from the table it is observed that as the first job is being processed, the other 4 jobs are waiting but all the 5 jobs are still in the system. When the first job is completed and delivered they are only 4 remaining in the system, for the next 17 days, 4 jobs are in the system, for days 22 to 35, 3 jobs are in the system, and so on there are 55 total days for the sequence. Therefore.

Average number of jobs in the system each day

$$= \frac{5(4)+4(17)+3(14)+2(9)+1(11)}{55}$$

= 2.89 jobs in the system/day.

(*d*) **Average job lateness.** This quantity gives an idea about the extent of lateness on an average for all jobs which were delivered after due date. The lateness for each job is the difference of its flow time and its due date. How time for job A, for example, is 4 days—since its due date is 6 days, the difference is 4–6, or –2 this means that job A was finished 2 days earlier than required and there was no lateness. Flow time for job B is 21 days, and its due date is 23 days and job is late by one day. Similarly, lateness for jobs C, D and E is 17, 32 and 43 days, respectively therefore,

$$\text{Average job lateness} = \frac{(0+1+17+32+43)}{5} = 186 \text{ days.}$$

(*ii*) **Sequencing of jobs using (shortest processing time) SPT Rule.** According to this rule, the highest priority is assigned to that job whose processing time is the shortest. The sequencing order and the corresponding data is shown in the table.

Waiting job (in SPT sequence)	Processing time (in days)	Flow time (in days)	Due date (in days from now)
A	4	4	6
D	9	13	12
E	11	24	12
C	14	38	18
B	17	55	20
Total	55		

(*a*) Total completion time = 55 days for all jobs to be completed.

(*b*) Average flow time $= \dfrac{4+13+24+38+55}{5} = 26.8$ days.

(*c*) Average number of jobs $= \dfrac{5(4)+4(9)+3(11)+2(14)+1(17)}{55}$

$= 2.44$ jobs per day.

(*d*) Average job lateness $= \dfrac{0+1+12+20+35}{5} = 13.6$ days.

II. Combined-Criteria Rules

If there are more than one criterion used for selecting the order in which jobs are to be processed, then they are known as combined-criteria rules or multiple-criteria rules. Some of the widely used combined-criteria rules are:

1. Johson's rule.
2. Critical ratio scheduling.
3. Index method of scheduling.

1. **Johnson's Rule.** Johnson's rule is used to determine the sequence of order for a series of jobs to be processed on a fixed number of machines. The basis for the sequencing is that the total time required to complete all the jobs should be minimum therefore reducing all idle times of all machines. There are several models in sequencing but the ones that are discussed in this book are:

Model 1. '*n*' jobs are to be processed through two machines say A and B in the order A, B only.

Model 2. '*n*' jobs are to be processed through three machines say A, B and C and in the order A, B, C only unless any other order is stated.

Model 3. '2' jobs are to be processed on '*n*' machines this is to be done by graphical method.

Model 4. '*n*' jobs '2' machines.

A set of steps known as the Johnson's rule are followed to find the optimum sequence of jobs. The steps are as follows:

Johnson's Rule for Scheduling

Step 1. Data showing the amount of time required for each job to be processed on each of the two machines are given. Among all the values, the shortest individual processing time is selected.

Step 2. If the shortest processing time is occurring on machine 1, then place the job as early as possible in the schedule. If the shortest time lies with machine-2, the job is placed as late as possible.

Step 3. Once a job has been scheduled, it is eliminated from further consideration, step 1 and step 2 are again applied to the remaining jobs. The scheduling sequence that is derived then applies to both machines and the resulting processing will given the minimum total elapsed time to complete all the jobs. The principal of Johnson's rule is that the least-time jobs are scheduled at the beginning and at the end of the sequence in order to maximize the amount of simultaneous operations on machines. 1 and 2, and so that the overall operating time to complete a specific number of jobs is minimized.

After finding the optimum sequences of jobs, the total minimum elapsed time to complete all jobs as well as the idle time for each machine is also found.

Worked Examples

1. A machine operator has to perform two operations, turning and threading on a number of different jobs. The times required to perform these operations are given. Determine the order in which the jobs should be processed in order to minimize the total time required to finish all the jobs. Also find the total minimum elapsed time and the idle time on each machine.

Job	M/C A Time for turning (hours)	M/C B Time for threading (hours)
1	3	8
2	12	10
3	5	9
4	2	6
5	9	3
6	11	1

Solution:

Application of Johnson rule to find the best sequence.

1. Among all the processing times, irrespective of which machine, the shortest processing time is selected. In the given table, the shortest time is for job 6 on machine B (1 min) so place the job as late as possible.

					6

2. The next shortest time is for job 4 on machine A(2 min.) So place the job as early as possible.

4					6

3. The next shortest time is a tie between jobs 1 and 5. Select any one place the job 1 as early as possible among the remaining slots.

4	1				6

4. The next shortest time is job 3 on machine A(5 min.) place it as early as possible.

4	1	3		5	6

5. The last remaining job 2 goes to the vacant slot.

4	1	3	2	5	6

∴ Best sequence for scheduling the jobs = 4 1 3 2 5 6

Job	Machine A	Machine B
4	2	6
1	3	8
3	5	9
2	12	10
5	9	3
6	11	1

To find total minimum elapsed time.

Row		Machine A		Machine B	
		In	Out	In	Out
(*i*)	4	0	2	2	8
(*ii*)	1	2	5	8	16
(*iii*)	3	5	10	16	25
(*iv*)	2	10	22	25	35
(*v*)	5	22	31	35	38
(*vi*)	6	31	42	42	43

Step to find total minimum elapsed time.

Row 1. Since job is selected as the first job in the sequence, it is processed straight away on machine A. Two hours later, it can directly enter machine B to be processed which takes 6 hours for the same. Job 1 is completed on machine B at the end of 8 hours.

Row 2. Job 1 is next in the sequence. It can be processed on machine A only after job 4 has moved from machine A to machine B. Job 1 takes 3 hours on machine A and comes out only at the end of 5th hour. But in order to be processed on machine B, job 1 has to wait for 3 hours since machine B is still busy with job 4. Job 1 can enter machine B at the end of 8th hours and needs 8 hours processing by the time the job is completed a cumulative of 16 hours have elapsed.

Row 3. Job 3 completes processing on machine A at the end of 10th hour but has to wait for 6 hours before being processed on machine B.

Rows 4, 5 and 6 are processed subsequently similar to the above steps. By the time job 6 completes processing on machine B, accumulative time of 43 hours would have elapsed.

∴ Total minimum elapsed time = 43 hours to complete all jobs.

To find idle items on machine A and machine B. To find this, the total duration of processing on machines A and B are found. In other words the duration for which the machines A and B are in process are added.

∴ Duration of processing on machine

$$A = 3 + 12 + 5 + 2 + 9 + 11 = 42 \text{ hours}$$

Duration of processing on machine

$$B = 8 + 10 + 9 + 6 + 3 + 1 = 37 \text{ hours.}$$

Now,

Idle time for machine A.

= Total minimum elapsed time – Duration of processing on machine A

= 43 – 42

Idle item for machine A = 1 hour

Idle time for machine B.

= Total minimum elapsed time – Duration of processing on machine B

= 43 – 37

Idle time for machine B = 6 hours.

2. A, B, C bakery company has orders for five speciality jobs (A, B, C, D and E) that must be processed sequentially through two work centres (Baking and Decoration). The amount of time (in hours) required for the job is shown below.

	Time required for processing (hours)				
Work centres	**A**	**B**	**C**	**D**	**E**
Baking (WC_1)	5	4	8	7	6
Decoration (WC_2)	3	9	2	4	10

Determine the schedule sequence that minimizes the total elapsed time for the five jobs, and present it in the form of a gantt chart.

Solution:

Applying Johnson's rule for sequencing, the jobs are scheduled one by one as follows:

Step 1.

				C

[∵ Processing time for job C on WC_2 is least]

Step 2.

			A	C

[∵ Processing time for job A on WC_2 is heart]

Step 3. [Tie between B and D, selected any one]

B			A	C

Setp 4.

B		D	A	C

[Processing time for job D on WC_2 is heart]

Setp 5. [∵ Last job is E]

B	E	D	A	C

∴ The best sequence for scheduling the jobs = BEDAC.

To find the total elapsed time using gantt chart best sequence table for 2 work centres are

JOB	B	E	D	A	C
WC_1 Bakery	4	6	7	5	8
WC_2 Decoration	9	10	4	3	2

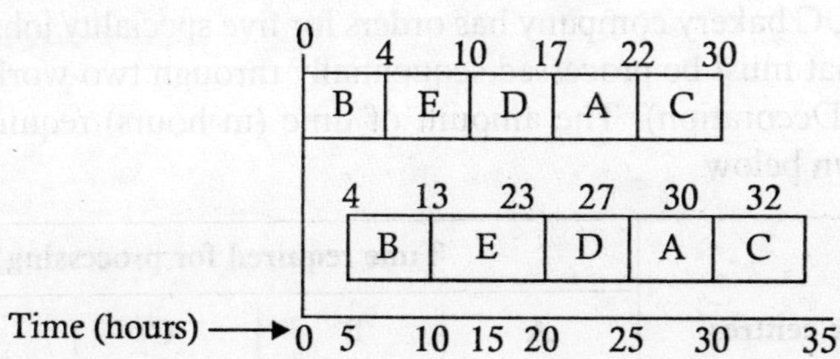

Model 2. '*n*' jobs '3' Machines. In these model '*n*' jobs are to be processed through 3 machines A, B, C in the order A, B, C, only unless any other order is stated.

In case of '*n*' jobs '3' machines there is no easy solution available. However, if a certain condition is satisfied. Model 2 problems can be converted into problems of model 1 type and then be solved. This is done in order to find only the best sequence while the remaining procedure remains the same as in the previous model.

Conditions to be satisfied to solve the problem.

The maximum processing time on mc/2 should be less than or equal to the minimum processing time on m/c 1.

Or

The maximum processing time on m/c 2 should be less than or equal to the minimum processing time on m/c 3.

Even if any one of the above condition is satisfied then the problem can be solved this is done by adding the processing times on machine 1 and machine 2 to make one column and then adding the processing times on machine 2 and machine 3 to make the second column. Now having only two columns, the best sequence is chosen by applying Johnson's rule as explained in model 1 of '*n*' jobs '2' machines.

Worked Problems

1. A machine operator has to perform three operations turning, threading and knurling on a number of jobs in that order.

Job	Machine 1 turning (min.)	Machine 2 threading (min.)	Machine 3 knurling (min.)
1	3	(8)	13
2	12	6	14
3	5	4	19
4	(2)	6	12
5	9	3	(8)
6	11	1	13

Solution:

Condition to be satisfied to solve the problem. First identify the maximum processing time on machine 2 and the minimum processing time on machine 1 and machine 3.

(*i*) Maximum processing time on machine 2 = 8.
(*ii*) Minimum processing time on machine 1 = 2.
(*iii*) Minimum processing time on machine 3 = 8.

The condition to be satisfied is that maximum processing time on m/c 2 should be less than or equal to either the minimum processing time on m/c 1 or the minimum processing time on m/c 3.

In the given table. (*i*) = (*iii*).

Therefore the problem can be solved.

Now, in the given table, add the processing time of each job on m/c 1 and m/c 2 to make one column, and then add the processing times of each job on m/c 2 and m/c 3 to make the second column.

Job	**m/c 1 + m/c 2 (min.)**	**m/c 1 + m/c 3 (min.)**
1	11	21
2	18	20
3	9	13
4	8	18
5	12	11
6	12	14

Now this problem has been temporarily converted into model 1 type *i.e.*, '*n*' jobs '2' machines. Now the best sequence is determined following Johnson's Algorithm.

Step. 1.

4					

∵ (Least element is 8 and is for job 4 in column 1)

2.

4	3				

∵ (Least element is 9 and is for job 3 in column 1)

3.

4	3	1			5

∵ (Least element is 12 and is a tie for job 1 in column 1 and for job 5 in column 2)

4.

4	3	1	6		5

∵ (Least element is 12 and is for a job 6 in column 1)

5.

4	3	1	6	2	5

∵ (Least job remaining is job 2)

∴ Best sequence

4	3	1	6	2	5

.

Now, the use of the second table is over. We go back to the original table with 3 columns to find the total minimum elapsed time and the idle times of each machine.

'Original Table' according to the best sequence.

Job	Machine 1 turning (min.)	Machine 2 threading (min.)	Machine 3 knurling (min.)
4	2	6	12
3	5	4	9
1	3	8	13
6	11	1	13
2	12	6	14
5	9	3	8
Total time	42	28	69

To find the total minimum elapsed time.

Total minimum elapsed time is found exactly the way it was done for model 1, except that there are 3 machines instead of 2.

		m/c 1		m/c 2		m/c 3	
Row	**Job**	**In**	**Out**	**In**	**Out**	**In**	**Out**
1	4	0	2	2	8	8	20
2	3	2	7	8	12	20	29
3	1	7	10	12	20	29	42
4	6	10	21	21	22	42	55
5	2	21	33	33	39	55	69
6	5	33	42	42	45	69	77

Explanation of the steps

Row 1. By the time job 4 comes out of m/c 3, 20 minutes would have elapsed.

Row 2. Job 3 has to wait for 1 min. and 8 min. at m/c 2 and 3 respectively before further processing continues.

Row 3. Job 1 has to wait again because m/c 2 and 3 were not ready yet.

Row 4. Job 6 completes its machining on m/c 1 on 21st minute now the m/c 2 which has already completed the earlier job (job 1) in the 20th minute has to wait for one minute before processing job. 6.

Row 5 and 6. As shown according to the table.

$\therefore$ Total minimum elapsed time = 77 min.

Idle time for m/c 1 = 77 – 42 = 35 min.

Idle time for m/c 2 = 77 – 28 = 49 min.

Idle time for m/c 3 = 77 – 69 = 8 min.

2. Find the sequence for the following 8 jobs, the minimizes the total elapsed time for completion of all jobs, each job being processed in the order CAB. Find the total elapsed time and idle time of each m/c.

Jobs	1	2	3	4	5	6	7	8
Machine A	4	6	7	4	5	3	6	2
B	8	10	7	8	11	8	9	13
C	5	6	2	3	4	9	15	11

The entries give the time in hours on the m/c.

Solution:

Table of jobs according to the machinery sequence.

Job	C	A	B
1	5	4	8
2	6	6	10
3	(2)	(7)	(7)
4	3	4	8
5	4	5	11
6	9	3	8
7	15	6	9
8	11	2	13

From the table it is seen that max. Time on m/c A is equal to the minimum time on m/c B. Hence the problem can be solved.

Job	C + A	A + B
1	9	12
2	12	16
3	9	14
4	7	12
5	9	16
6	12	11
7	21	15
8	13	15

Best sequence of jobs by applying Johnson's rule.

4	5	3	1	2	8	7	6

Note. The above sequence is obtained by selecting the jobs in the order 4, 5, 3, 1, 6, 8 and 7.

Therefore table according to the best sequence is:

Job	C	A	B
4	3	4	8
5	4	5	11
3	2	7	7
1	5	4	8
2	6	6	10
8	11	2	13
7	15	6	9
6	9	3	8
Total duration for each m/c	55	37	74

To find total minimum elapsed time.

Job	C		A		B	
	In	Out	In	Out	In	Out
4	0	3	3	7	7	15
5	3	7	7	12	15	26
3	7	9	12	19	26	33
1	9	14	19	23	33	41
2	14	21	23	29	41	51
8	20	31	31	33	51	64
7	31	46	46	52	64	73
6	46	54	54	57	73	81

Therefore, total minimum elapsed time = 81 hrs.

Idle time for m/c A = 81 – 37 = 44 hrs.

Idle time for m/c B = 81 – 74 – 07 hrs.

Idle time for m/c C = 81 – 55 = 26 hrs.

Model 3. '2' jobs '*n*' Machines. The previous 2 models had 2 or 3 machines with the possibility of accommodating any number of jobs. In this model, there are 2 jobs which needs processing on many machines. The decision to be taken here is as to which job should be taken up first by each of the machines specified. In other words, each machine processes on of the two jobs. First and the other job next in order to see that the total time elapsed to complete both jobs are least.

Johnson's rule cannot be used for model 3 but a graphical method is made use to solve this problem.

The procedure to solve problems in model 3 by graphical method are:

Step 1. Draw two axes at right angles to each other represent the processing time of job 1 along the horizontal axis and processing time of job 2 along the vertical axis.

Step 2. Indicate the machining times for the two jobs on corresponding axes in the given processing order for each job. In other words, along *x*-axis indicate the machining sequence of job 1 and mark off the corresponding machining times. Repeat the same for job 2 along *y*-axis.

Setp 3. Rectangular blocks are drawn by matching the same machines on the *x* and *y* axis.

Step 4. Draw a path starting from the origin and at a angle of 45° and move towards the point of finish. If blocks are encountered draw either horizontal lines or vertical lines to pass over to blocks. A horizontal line represent job 1 being in process while job 2 remains idle: A vertical line represent work on job 2 is going on while job 1 is remaining idle; and the 45° line represents simultaneous work on both jobs but on different machines.

Step 5. Many paths can be drawn from start to finish. The answer to the scheduling path is solved by selecting one of these paths. The path is actually selected whose sum of horizontal and vertical portions are the least among all the paths. In other words, that path is selected whose job is to be done first on each machine by observing the horizontal vertical and 45° lines.

45° line indicates = simultaneous work of both jobs on machines as shown by *x*-axis and *y*-axis.

Horizontal line indicates = J_1 is in process J_2 is idle.

Vertical line indicates = J_2 is in process J_1 in idle.

Resource Allocation Methods

The building blocks of the corporate resource allocation strategy are the following elementary investment strategies.

1. Replacement and modernisation.
2. Capacity expansion.
3. Vertical integration.
4. Concentric diversification.
5. Conglomerate diversification.
6. Divestment.

Replacement and Modernisation

This is often regarded as an integrate part of the base level capital expenditure program. Replacement and modernisation investments are meant to maintain the production capacity of the firm, improve quality and reduce costs. Without such investments, which are undertaken more or less rountinely by well managed firms, the competitive strength of the firm in its existing line of business can be significantly impaired. If such investments are neglected, the very existence of the firm can be jeopardised.

E.g. Textile Industries.

Capacity Expansion

When a company anticipates growth in market size of its product range or increase in the market share enjoyed by it in its product range, expansion of the capacity of the existing product range would have a great approach. Such an expansion offers several advantages:

(*a*) Familiarity with technology.

(*b*) Production methods.

(*c*) Market conditions.

(*d*) Lower capital costs due to existence of surplus capacity in certain sections of the industry.

(*e*) Reduction in unit overhead costs due to larger volumes of production.

Vertical Integration

This may be of two types. Backward integration and forward integration.

Backward Integration. This involves the manufacture of raw materials and the components required for the existing operation of the company. *E.g.*, Reliance Industries Limited.

Forward Integration. This involves the manufacture of products that use the existing products of the company as input,

E.g. Bharat forge company, setup an automotive. Axle unit that uses its forgings as an input.

Vertical integration provides greater stability to the existing as well as proposed operation, brings about better co-ordination between the supplying and consuming units and enhances the market power of the combined entity.

Concentric Diversification

Many companies seek to widen their product range by adding related products. *E.g.* A soap manufacturer may enter the field of detergents. A scooter producer may add motor cycles to its product lines. The advantages of such diversification, referred to a concentric diversification or allied diversification are.

1. Company can economise in procurement of raw materials.
2. Certain common manufacturing facilities may be utilised.
3. All products may benefit from the market image of the company.
4. A common distribution network may be employed.

Conglomerate Diversification

It involves investment in fields unrelated to the existing line of business. *E.g.* an engineering company like Larsen and Toubro invests in shipping. It is a case of conglomerate diversification. The principle motivations underlying the conglomerate diversification are:

(*a*) To overcome the limited growth opportunities in the existing line of business.

(*b*) To reduce the overall risk exposure of the firm.

Divestment

Divestment is the opposite of investment. This involves termination or liquidation of the plant or even a division of a firm. The disposal of the chembur plant of Union Carbide to Oswal Agro is an example of divestment. The motivation for divestment may be:

(*a*) Low profitability.

(*b*) Declining market share.

(*c*) Difficulty in managing.

(*d*) Incompatibility with corporate strategy.

Work-Breakdown Structure

The WBS as its name suggests represents a systematic and logical breakdown of the project into its component and parts and is constructed by dividing the project into its major parts with each of these being divided into sub parts. This continued till a breakdown is done in terms of manageable units of work for which responsibility can be defined. Thus the work breakdown structure helps in

1. Effective planning by dividing the work into manageable elements which can be planned budgeted and controlled.
2. Assignment of responsibility for work elements to project personnel and outside agencies.
3. Development of control and information system.

The WBS is the single most important element because it provides a common framework from which.

- ✓ The total program can be described as a summation of subdivided elements.
- ✓ Planning can be performed.
- ✓ Cost and budgets can be established.
- ✓ Time cost and performance can be tracked.
- ✓ Scheduling and status reporting procedures can be established.
- ✓ The responsibilities of each element can be established.

Work breakdown structure and project organisation

The project organisation represents formally how the project personnel and outside agencies are going to work. The WBS defines what work is done in a detailed manner. To assign responsibility for the task to be done. The work breakdown structure has to be integrated with the project organisation structure. This calls for blending the vertical breakdown of the work (as arrived in the work breakdown structure) with the project organisation structure. This results in deligation of the project task which are the specific responsibilities of organisational units/managers.

WORK BREAKDOWN STRUCTURE (WBS)

The breaking down of large activities into comprehensible or manageable units is a fundamental part of project management. Figure shows how a systems project "the installation of a new computer system" was broken down into elements that one person or one department could tackle as an activity in its own right.

WBS is also known as 'chunking' or 'unbundling'. This is attractive as it gives people responsibility for a manageable part of the project. WBS also facilitates financial control activities, as individual parts can have their consumption of resources tracked. The sensitivity of such control is preferable to keeping track of a single large activity.

The major drawback is that people naturally try to optimise their own part of the project and lose the holistic view of the project, positions can become entrenched and destructive when improvements in one area upset the running of another, *e.g.*, introducting new accounting procedures may greatly. Streamline the running of the accounting departments role in projects, but if it doubles the administrative workload for another function, the result is that neither the project nor the organisation as a whole has benefited.

The role of WBS is to create a linked, hierarchial series of activities which are independent units but at the same time still part of the whole. The hierarchial structure can be represented using the input, output, constraints and mechanisms model from chapter 1.

Work Breakdown Structure

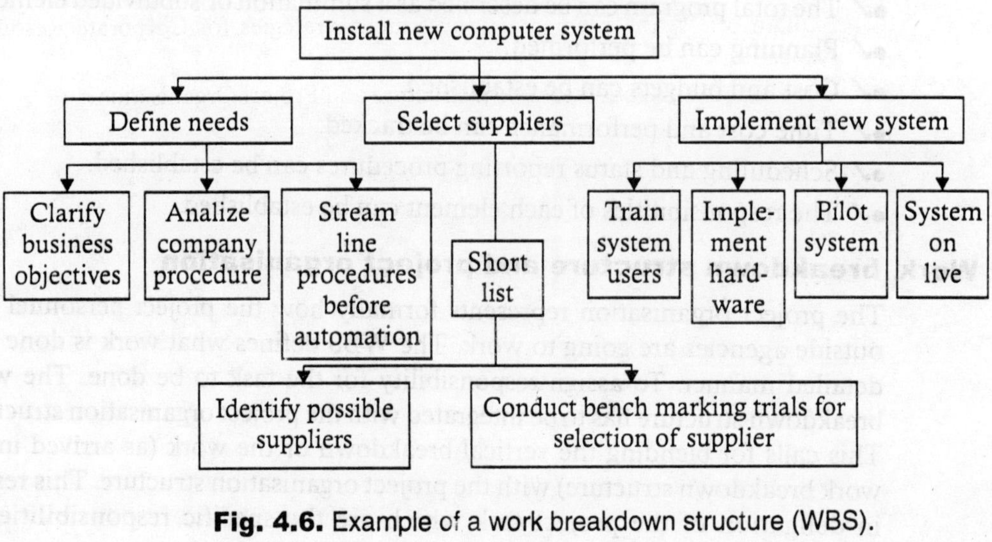

Fig. 4.6. Example of a work breakdown structure (WBS).

This form of activity model has been used effectively for a number of years, having been originally developed by the U.S Air force. Models are created at a number of levels which are essentially static representations of the reality of project situations, *i.e.,* they are a true representation of reality at one point in time.

The same technique can be applied very effectively in line-management situations where previously the description of procedures would have to be written long hand. Process mapping techniques, such as those discussed throughout this book, work far better and greatly cut down the amount of documentation required, while improving the usefulness of the end results.

Review Questions

1. Discuss briefly Project Implementation Schedule.
2. Illustrate the various factors in Time-Setting.
3. Explain how a schedule for feasibility study is prepared?
4. What are the techniques followed to have "Effective Time Management"?
5. Define Scheduling. Discuss the objectives of Scheduling.
6. List out some of the Scheduling Strategies.
7. What is Forward and Backward Scheduling?
8. Distinguish between Forward and Backward Scheduling.
9. Explain the different guidelines to be followed in Scheduling.
10. Classify the Scheduling methods.
11. Discuss the importance of Gautt Chart in Scheduling.
12. Explain briefly the Elementary investment strategies for Corporate Resource Allocation.
13. Write a note on Work breakdown structure and Project Organisation.

5 TOOLS AND TECHNIQUES OF PROJECT MANAGEMENT

INTRODUCTION

The tools and techniques enable effective communication between the various members involved in the project as they serve as the common language that is understood by all recipients. The basic idea of using tools is to give a clear picture of the progress of any event including a project. The main parameters that are usually shown using the tools are:

Cost.

Time.

Performance.

There are a variety of tools that are being used to represent various activities. The exact tool that is selected from this large number must depend upon the ultimate receiver of information for *e.g.* : the top management will not be interested in details of each activity but they will only be interested in knowing how the final picture is represented in terms of cost and schedule, whereas the lower level employees will need information in as much detail as possible of the large number of tools available, one of the most commonly used one is the — Gantt Chart which is basically a bar chart.

Gantt Chart (Bar Chart)

Gantt Chart. This is one of the most common type of display and is named after **Henry Gantt** a french man who first used in the early 1900's. The bar chart is a means of displaying simple activities against time or money. An activity represents the amount of work required to proceed from one point to another. Bar charts are used for indicating program progress and usually include activity duration, schedules date, etc.

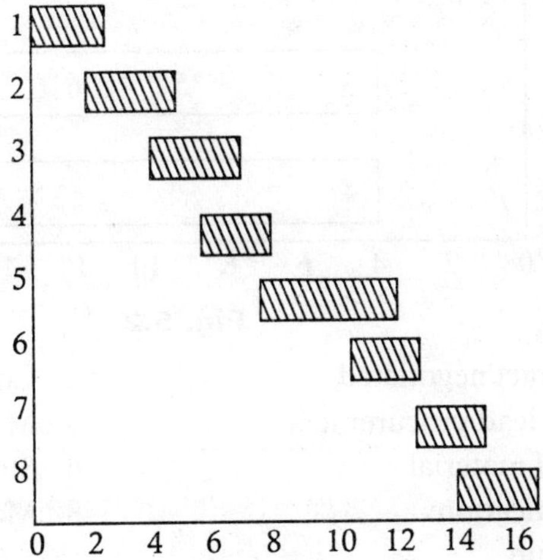

Fig. 5.1

The figure shows the different activities that are needed to start the production line for a new product. Each bar represents the start and completion of each activity and the time required for each activity from start to end.

Advantages

- Simple to understand and easy to change.
- This is the simplest and least complex method to portray progress or the lack of progress.
- The chart can be expanded to identify certain elements, that may be ahead or behind schedule.

Disadvantages

- Bar charts do not show the interdependency of activities which is a very important parameter to control project cost as this feature is missing, the bar chart has little predictive value.
- Bar chart cannot show the results of a late or early start in activities.
- Bar charts do not show the uncertainty involved in completion of that activity and is thus insensitive.

The above drawing shows the bar chart for a single activity. A bar chart for combined activity is as shown below:

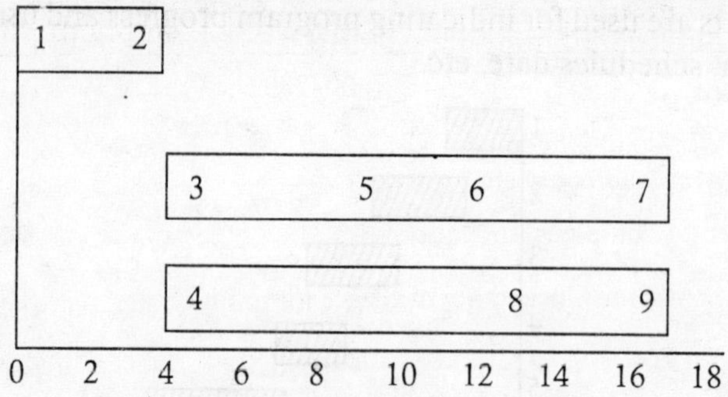

Fig. 5.2

1. Contract negotiated
2. Contract signed
3. Long lead procurement
4. Manufacturing schedule
5. Bill of material
6. Short lead procurement
7. Specifications
8. Manufacturing plans
9. Start up.

The bar chart for combined activities can be used to overcome a few disadvantages of the single activity bar chart. But this also has an inherent weakness, which is, the number do not indicate-clearly the start or end of the activity. Thus the numbers must represent events rather than activities. Another drawback of this chart is that it fails to clearly define the relationship between the activities on the same bar. For example, must even 3 be completed before event 5 is started? This can be overcome by placing milestone or small triangles at strategic locations on the bar to indicate the starting and completion of the event.

Application or Example

Gantt Chart For Scheduling Daily Work over a Period of 12 Days

The Fig. 5.3 shows the conventional work schedule in production facility for 12 days. The work begins on Monday and proceeds unhindered till Thursday when the work is stopped due to lack of material. The work again proceeds unhindered on Friday. But due to stoppage, the total production till Friday is only 280 as against the planned production of 300. The production line was not available on Saturday and Sunday. Work resumed on Monday but was stopped on Tuesday as the machine was down for repairs work started only on Thursday but the operations were sporadic and output was poor. Thus the total production for the week was only 340 as against the planned production of 400.

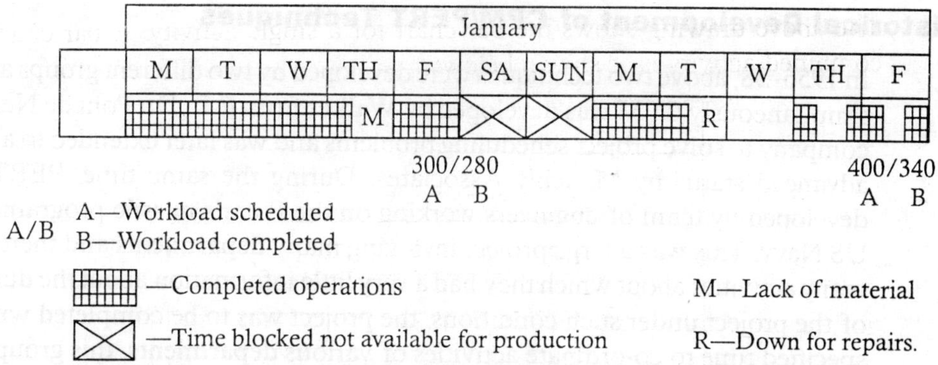

Fig. 5.3

Logic Diagrams and Networks

Logic diagrams are usually the tougher to draw. These are developed to illustrate the inductive and deductive reasoning necessary to achieve some objective within a time frame. The major difficulty in developing logic diagrams is having to answer key questions like what happens if something goes wrong? etc. logic diagrams are constructed similar to bar charts on the assumption that nothing will go wrong and are usually accompanied by detail questioning possibly in a check list format, which require answering.

The logic diagram for showing the logic flow for production of a piece X is as shown:

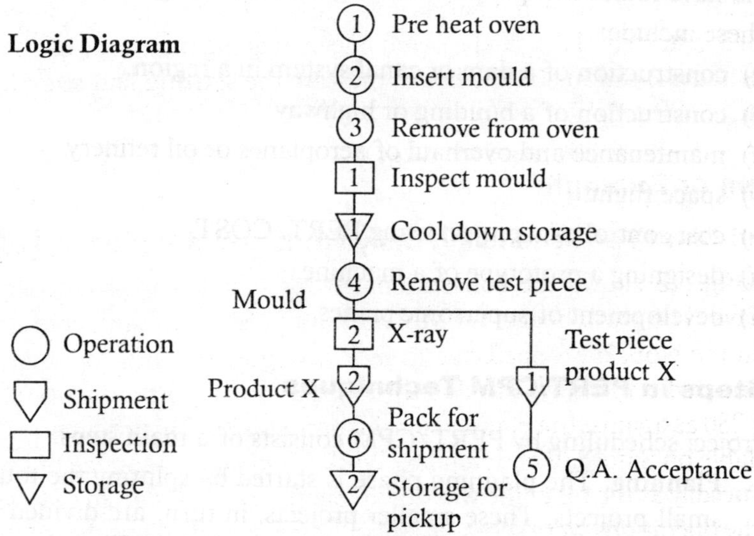

Fig. 5.4

Historical Development of CPM/PERT Techniques

In 1956–58, above two techniques were developed by two different groups almost simultaneously CPM was developed by **Walker** from E.L. Du Pont de Nenours company to solve project scheduling problems and was later extended to a more advanced status by Mauchly Associates. During the same time, PERT was developed by team of engineers working on the Polar's missile programme of US Navy. This was a large project involving many departments and there were many activities about which they had a very little information about the duration of the project under such conditions, the project was to be completed within a specified time to co-ordinate activities of various departments, this group used PERT and devised the technique independent of CPM.

The methods are essentially network-oriented techniques using the same principal PERT and CPM are basically time oriented methods in the sense that they both lead to the determination of a time schedule for the project. The significant difference between two approaches is that the time estimates for the different activities in CPM were assumed to be deterministic while is PERT these were described probabilistically. Nowadays, PERT and CPM actually comprise one technique and the differences, if any are only historical. Therefore, these techniques are referred as a 'project scheduling' techniques.

Applications of PERT/CPM Techniques

These methods have been applied to a wide variety of problems in industries and have found acceptance even in government organisations.

These include:

(*i*) construction of a dam or canal system in a region.

(*ii*) construction of a building or highway.

(*iii*) maintenance and overhaul of aeroplanes or oil refinery.

(*iv*) space flight.

(*v*) cost control of a project using PERT/COST.

(*vi*) designing a prototype of a machine.

(*vii*) development of supersonic planes.

Basic Steps in PERT/CPM Techniques

Project scheduling by PERT/CPM consists of a main steps:

1. **Planning.** The planning phase is started by splitting the total project into small projects. These smaller projects, in turn, are divided into activities and are analysed by the department or a section. The relationship of each

activity with respect to other activities are defined and established, and the corresponding responsibilities and the authority are also stated. Thus, the possibility of overlooking and talk necessary for the completion of the project is reduced substantially.

2. **Scheduling.** The ultimate objective of a scheduling phase is to prepare a time chart showing the start and finish times for each activity as well as its relationship to other activities of the project. Moreover, the schedule must pinpoint the critical path (in view of time) activities which require special attention if the project is to be completed in time for non-critical activities, the schedule must show the amount of slack or float times (defined later) which can be used advantageously when such activities are delayed or when limited resources are to be utilised effectively in this phase, it is possible to resource requirements such as time, manpower, money, machines, etc.
3. **Allocation of resources.** Allocation of resources is performed to achieve the desired objective. A resource is a physical variable such as labour, finance, equipment and space which will impose a limitation on time for the project, when resources are limited and conflicting, demands are made for the same type of resources a systematic method for allocation of resources become essential resource allocation usually incurs a compromise, and the choice of this compromise depends on the judgement of managers.
4. **Controlling.** The final phase in project management is controlling. Critical path methods facilitate the application of the principle of management by expectation to identifying areas that are critical to the completion of the project. By having progress reports from time to time and updating the network continuously, a better financial as well as technical control over the project is exercised. Arrow diagram and time charts are used for making periodic progress reports. If necessary, new course of action is determined for the remaining portion of project.

Network Diagram Representation

In project scheduling, the first step is to sketch an arrow diagram which shows inter-dependencies and the precedence relationship among activities (as defined below) of the project in a network representation of a project, certain basic definitions are used:

1. **Activity.** *Any individual operation, which utilises resources and has an end and a beginning, is called activity.* An arrow is commonly used to represent an activity with its head indicating the direction of progress in the project. These are usually classified into following categories:

(*i*) ***Predecessor activity.*** Activities that must be completed immediately prior to the start of another activity are called predecessor activities.

(*ii*) ***Successor activity.*** Activities that cannot be started until one of more of other activities are completed but immediately succeed than are called successor activity.

(*iii*) ***Concurrent activity.*** Activities which can be accomplished concurrently are known as concurrent activities it may be noted than an activity can be a predecessor or a successor to an event or it may be concurrent with one or more of the other activities.

(*iv*) ***Dummy activity.*** An activity which does not consume any kind of resource but merely depicts the technological dependence is called a dummy activity. It may be noted that the dummy activity is inserted in the network to classify the activity pattern in the following two situations:

(*a*) to make activities with common starting and finishing points distinguishable, and

(*b*) to identify and maintain the proper precedence relationship between activities that are not connected by events.

For example, consider a situation where A and B are concurrent activities, C is dependent on A, and D is dependent on A and B both such a situation can be handled by using a dummy activity as follows (Fig. 5.5).

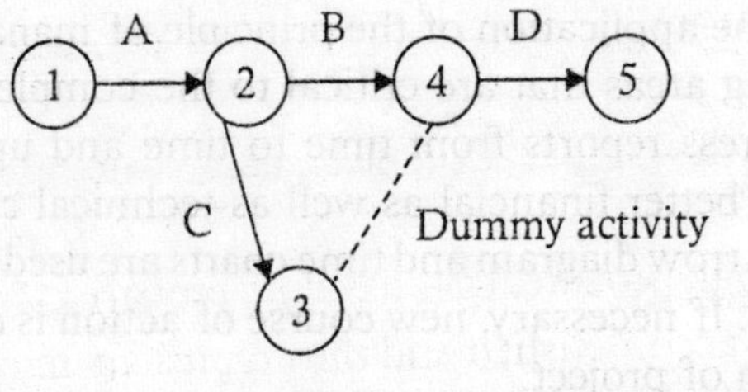

Fig. 5.5

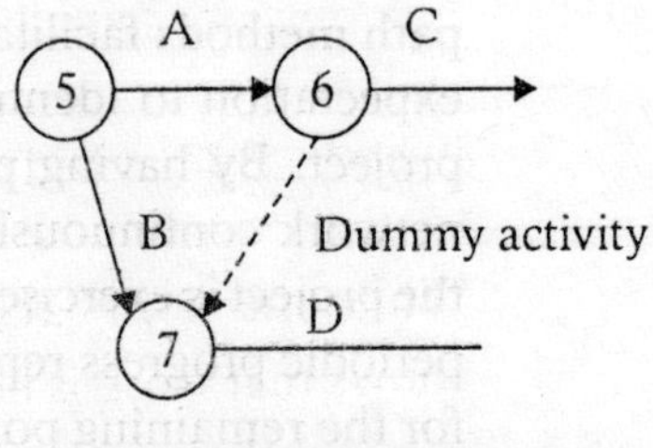

Fig. 5.6

In another situation, consider the following diagram where job B and C have the same job reference and they can be started independently on completion of A. But, D could be started only after completion of B and C. This relationship is shown by the dotted line. (Fig. 5.6).

2. **Event.** An event represents a point in time signifying the completion of some activities and the beginning of new ones. This is usually represented by a circle 'O' in a network which is also called a node or connector.

The events can be further classified into following three categories (as shown below in the Fig. 5.7).

(*i*) ***Merge event.*** When more than one activity comes and joins an event, such event is known as merge event.

(*ii*) ***Burst event.*** When more than one activity leaves an event, such event is known as a burst event.

(*iii*) ***Merge and burst event.*** An activity may be merge and burst event at the same time as with respect to some activities it can be a merge event and with respect to some other activities it may be a burst event.

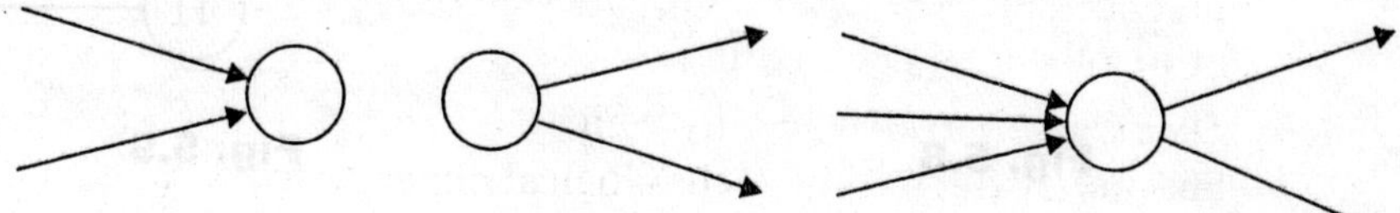

Merge event Burst event Merge and burst event

Fig. 5.7

Remarks

1. An event is that particular instant of time at which some specific part of a project has been or is to be achieved. While, an activity is actual performance of a task. An activity requires time and resources for its completion.
   Examples of events: design completed, pipeline laid, etc.
   Examples of activities: assembly of parts, mixing of concrete, preparing budget, etc.
2. Events are described by such words as: Complete, start, issue, approve, tested etc.
   While, the word like: design procure, test, develop, prepare etc.
   Shows that work is being accomplished and thus represent activities.
3. While drawing networks, it is assumed that:
   (*i*) time flows from left to right, and
   (*ii*) head events always have number higher than that of tail event. Thus activity (*i–j*) always means that the job which begin at event '*i*' is completed at event *j*.
4. Network representations is based on the following two axioms:
   (*i*) An event is not said to be complete until all the activities flowing into it are completed.
   (*ii*) No subsequent activity can begin until its tail event is reached or completed.

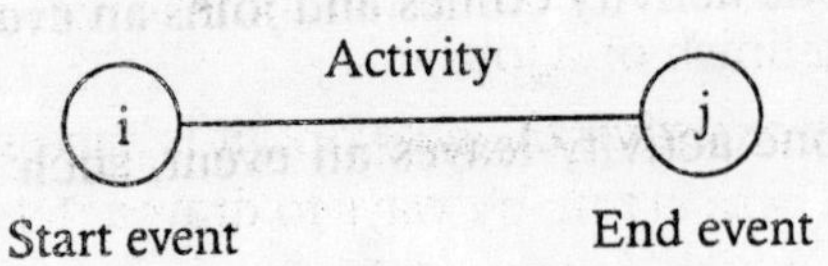

Fig. 5.8

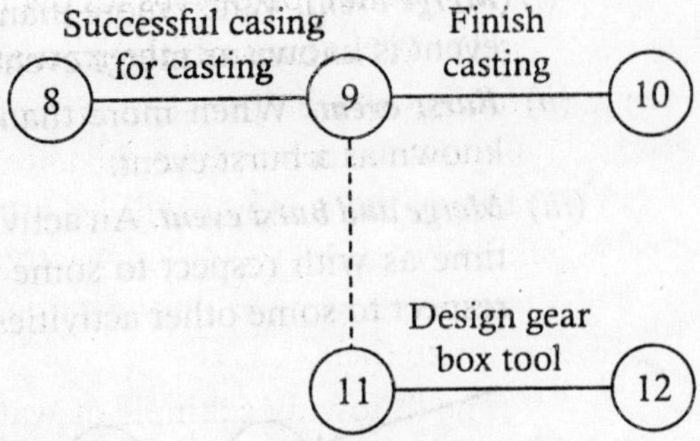

Fig. 5.9

Illustration designing tool for a gear box is an activity. A decision to start designing tools may depend on having a successful casing for gear box casting.

In terms of technological sequence, casting as such has little bearing on the tooling of the gear box, but the management would prefer to have a successful casing before gear-box tool is designed.

Thus the dependence of gear-box tools on successful casing is shown as a dummy activity (Fig. 5.9).

It is also important to note that the length of the arc (or arrow) need not be proportional to the duration of the activity nor does it have to be drawn as a straight line.

If the duration of each activity as well as their logical sequence is known, it can be shown on a network (Fig. 5.10) the duration may be measured in days.

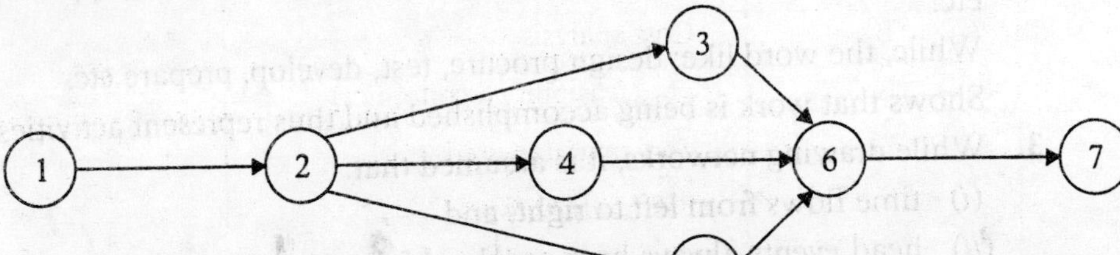

Fig. 5.10. Activity duration on the network.

5. **Sequencing.** The first prerequisite in the development of a network is to maintain the precedence relationship. In order to make a network, following points should be taken into consideration:
   (*i*) What job or jobs precede it?
   (*ii*) What job or jobs could run concurrently?

(*iii*) What job or jobs follow it?
(*iv*) What controls the start and finish of a job?

Since all further calculations are based on the network, it is necessary that a network be drawn with full care. There are many ways to draw a network. In this text, the method will be used which follows the precedence table. It is generally agreed the dummy activities be used as liberally as needed. In the first attempt, while revising the same network, every attempt should be made to minimize then.

The following example of water pump is to be designed for an automobile major specifications are given.

Following list represents major activities for effective control of the project:

1. Drawing prepaid and approved
2. Cost analysis
3. Tools feasibility (economics)
4. Tool manufactured
5. Favourable cost
6. Raw materials procured
7. Sub-assemblies ordered
8. Sub-assemblies received
9. Parts manufactured
10. Final assembly
11. Testing and shipment.

Based on this available information, a precedence table may be formed (Table 5.1).

Precedence Table 5.1

Activity	Description	Preceding Activity
A	Drawing prepared and approved	—
B	Cost analysis	A
C	Tool feasibility (economics)	A
D	Tool manufactured	C
E	Favourable cost	B, C
F	Raw materials procured	D, E
G	Sub-assemblies ordered	E
H	Sub-assemblies received	G
I	Parts manufactured	D, F
J	Final assembly	I, H
K	Testing and shipment	J

In this table, due consideration has been given to precedings of an activity/ activities. While drawing the network, other factors will be considered.

Step 1. The activity 'A' has no preceding activity and is represented by an arrowed line (Fig. 5.11).

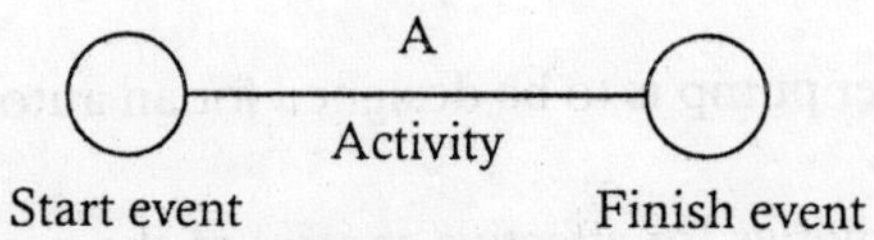

Fig. 5.11

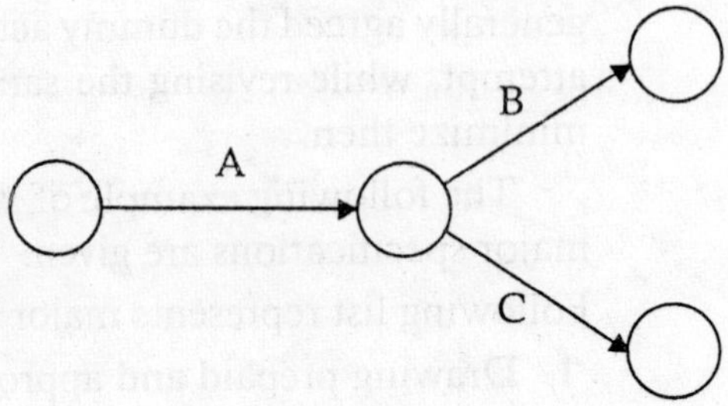

Fig. 5.12

Step 2. Activities B and C are preceded by an activity 'A' and activities B and C could be done concurrently (if resources are not binding) no other activity can be scheduled at this stage. This is shown in Fig. 5.12.

Step 3. The activity 'D' can be sequenced easily. Cost favourable activity 'E' cannot be scheduled unless activities B and C are scheduled. Further, it is observed that dependence of 'cost favourable activity' on the 'Economy of tooling' is from a 'Technical view-point' and does not consume any resources, and hence this dummy activity should be shown by a dotted line. Thus, the network up to this stage is shown in Fig. 5.13.

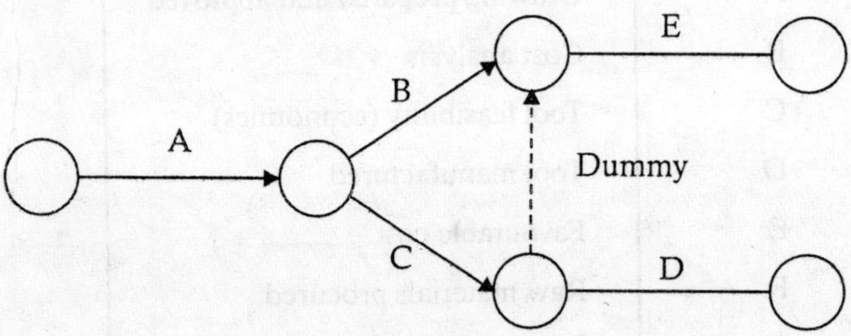

Fig. 5.13

Rest of the network is straight forward, and thus the complete network is shown in Fig. 5.14.

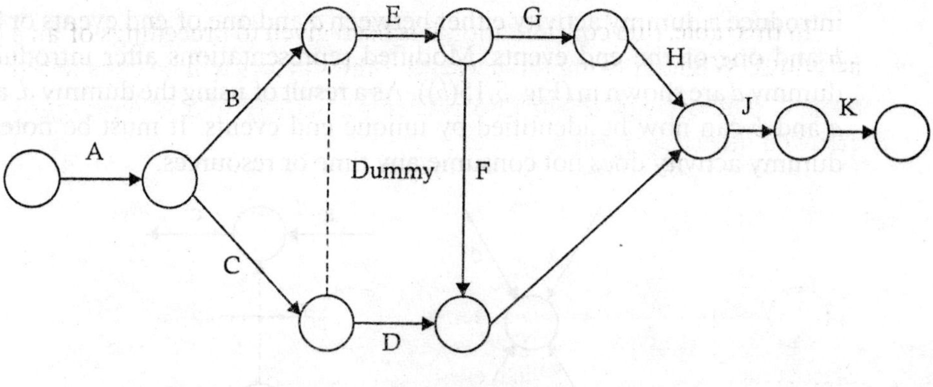

Fig. 5.14

Rules for Drawing Network Diagram

Rules for drawing network diagrams are summarized as follows:

Rule 1. Each activity is represented by one and only one arrow in the network.

This implies that no single activity can be represented twice in the network. This is to be distinguished from the case where one activity is broken into segments, in such a case, each segment may be represented by a separate arrow. For example in laying down a pipe, this may be done in sections rather than as one job.

Rule 2. No two activities can be identified by the same end events. For example, activities *a* and *b* (Fig. 5.15(*a*)) have the same end events. The procedure is to

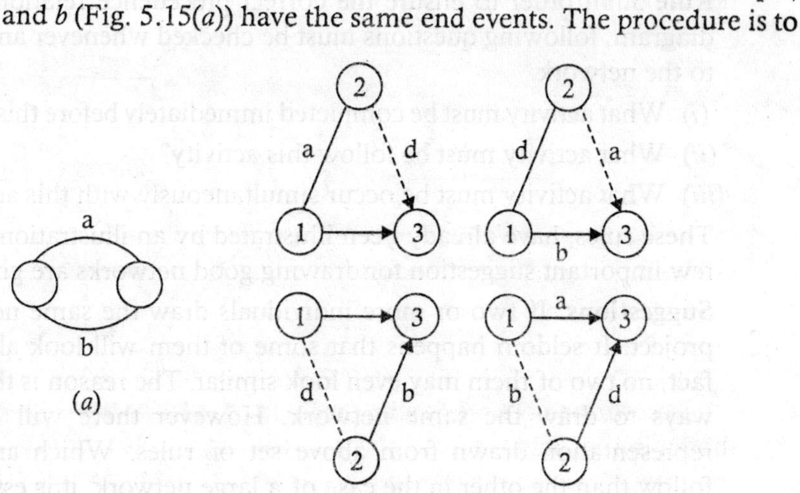

Fig. 5.15

introduce a dummy activity either between *a* and one of end events or between *b* and one of the end events. Modified representations after introducing the dummy *d* are shown in (Fig. 5.15(*b*)). As a result of using the dummy *d*, activities *a* and *b* can now be identified by unique end events. It must be noted that a dummy activity does not consume any time or resources.

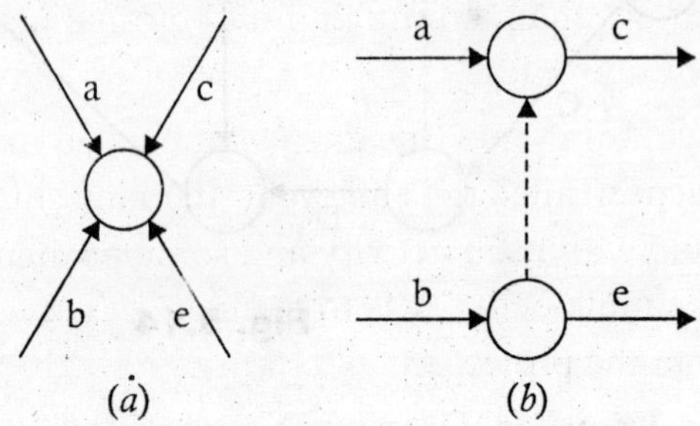

Fig. 5.16

Dummy activities are also useful in establishing logic relationship in the allow diagram which otherwise cannot be represented correctly. Suppose jobs *a* and *b* in a certain project must precede the job *c*. On the other hand, the job *e* is preceded by the job *b* only (Fig. 5.16(*a*)) shows the incorrect way since, through the relationship between *a*, *b* and *c* are correct, the diagram implies that the job must be preceded by both the jobs *a* and *b*. The correct representation using the dummy *d* is shown obvious that indicated precedence relationship are justified.

Rule 3. In order to ensure the correct precedence relationship in the arrow diagram, following questions must be checked whenever any activity is added to the network.

(*i*) What activity must be completed immediately before this activity can start ?

(*ii*) What activity must be follow this activity?

(*iii*) What activity must be occur simultaneously with this activity?

These rules, have already been illustrated by an illustration apart from this, a few important suggestion for drawing good networks are given below:

Suggestions. If two or more individuals draw the same network for a given project. It seldom happens that some of them will look alike as a matter of fact, no two of them may even look similar. The reason is that there are many ways to draw the same network. However there will be same network representation drawn from above set of rules. Which are much easier to follow than the other in the case of a large network, it is essential that certain 'good habits' be practised to draw an "easy to follow" network.

(*i*) Try to avoid arrows which cross each other.

(*ii*) Use straight arrows.

(*iii*) Do not attempt to represent duration of activity by its arrow length.

(*iv*) Use arrow from left to right (or right to left) avoid mixing two directions vertical and standing arrows may be used if necessary.

(*v*) Use dummies freely in rough draft but final network should not have any redundant dummies.

(*vi*) The network has only one entry point called the start event and one point of emergence-called the end event. In many situations all these may not be compatible with each activity and some of them are violated. The idea of having a simple network is to facilitate easy reading for all those who are involved in the project.

6. Common Errors In Drawing Networks. Three types of errors are most commonly observed in drawing network diagrams :

(*i*) ***Dangling.*** To disconnect an activity before the completion of all activities in a network diagram is known as dangling as shown in Fig. 5.17, activities (5–10) and (6–7) are not the last activities in the network. So the diagram is wrong and indicates the error of dangling.

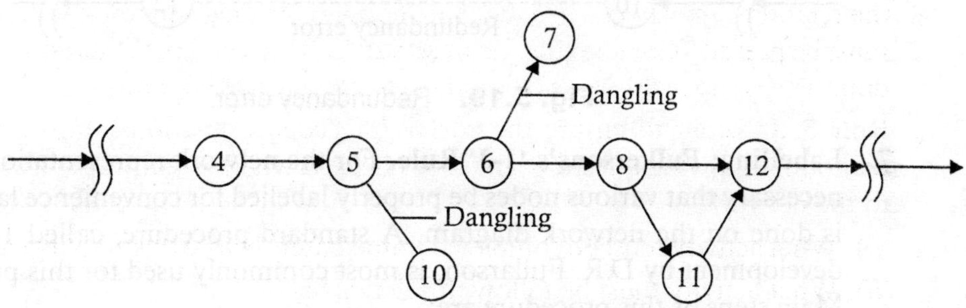

Fig. 5.17. Dangling error diagram.

(*ii*) ***Looping (or cycling).*** Looping error is also known as cycling error in a network diagram. Drawing and endless loop in a network is known as an error of looping as shown in the Fig. 5.18.

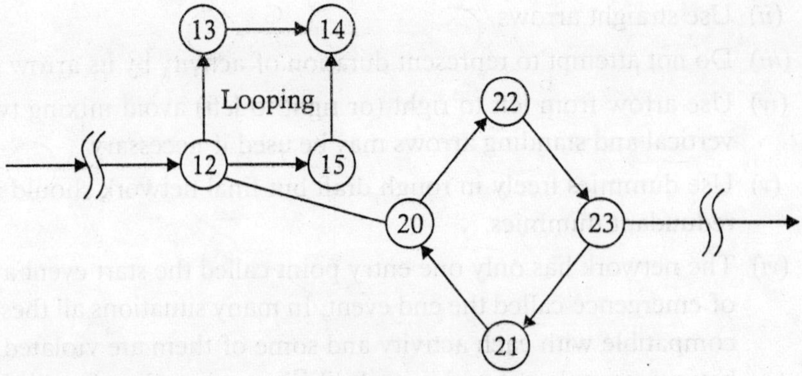

Fig. 5.18. Looping (or cycling) error diagram.

(*iii*) ***Redundancy.*** Unnecessarily inserting the dummy activity in a network logic is known as the error of redundancy as shown in Fig. 5.19.

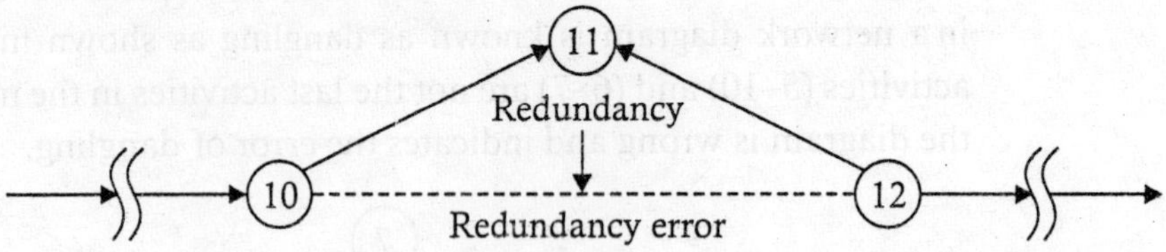

Fig. 5.19. Redundancy error.

7. **Labelling: Fulkersons's '1-J' Rule.** For the network representations, it is necessary that various nodes be properly labelled for convenience labelling is done on the network diagram. A standard procedure, called 1-J' rule development by D.R. Fullarson, is most commonly used for this purpose. Main steps of this procedure are:
   (*a*) A start event is the one which has arrows emerging from it but none entering it. Find the start event and number it as unit (1).
   (*b*) Delete all arrows emerging from all numbered events. This will create at least one new start event out of preceding event.
   (*c*) No all new start events '2', '3', and so on (no definite rule is necessary; but numbering from 'top to bottom' may facilitate other users in reading the network when there are more than one new start events.
   (*d*) Go on repeating steps number (*b*) and (*c*) until the end is reached now consider the network diagram of (Fig. 5.20) for 'Labelling' by Fulkerson's rule.

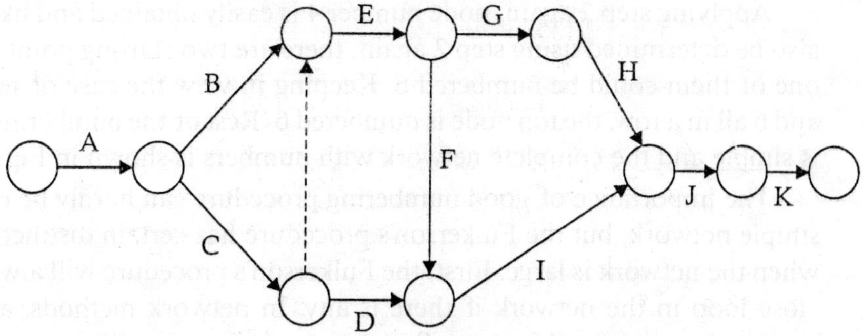

Fig. 5.20

To number the nodes using Fullerson's rule, numbering of nodes 1 and 2 is obvious. Apply step 2. The bottom node is the only node from jobs which is emerging out but none extering it. This is number 3.

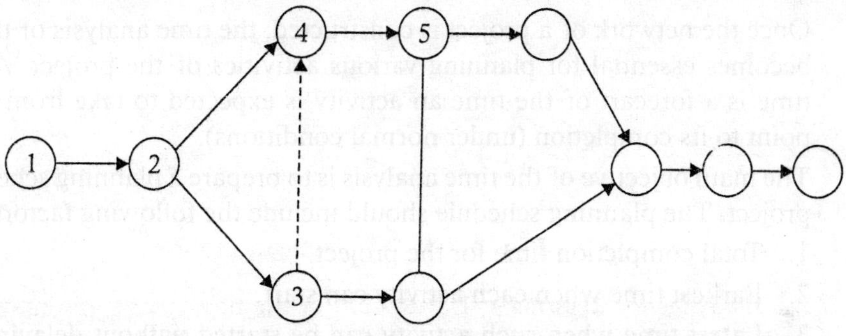

Fig. 5.21

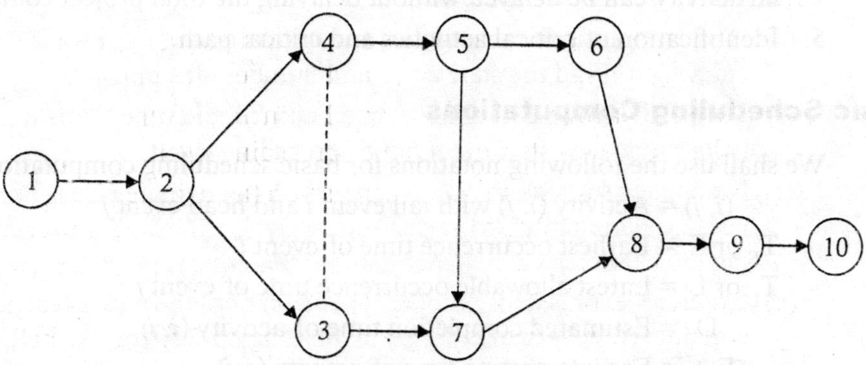

Fig. 5.22

Applying step 2 again, node number 4 is easily obtained and likewise 5 can also be determined using step 2 again, there are two starting points, and either one of them could be numbered 6. Keeping in view the case of numbers 4, 5 and 6 all in a row, the top node is numbered 6. Rest of the numbering procedure is simple and the complete network with numbers is shown in Fig. 5.22.

The importance of good numbering procedure can hardly be claimed in a simple network, but the Fulkerson's procedure has certain distinct advantages when the network is large. First, the Fulkerson's procedure will always detect a close loop in the network if there is any. In network methods, a close loop represents an impossible event. Second, numbers are smaller towards the start, side and become larger on the end.

A third advantage will become apparent when a matrix representation of the network is bought for computerization.

Time Estimates and Critical Path in Network Analysis

Once the network of a project is constructed, the time analysis of the network becomes essential for planning various activities of the project. An activity-time is a forecart of the time an activity is expected to take from its starting point to its completion (under normal conditions).

The main objective of the time analysis is to prepare a planning schedule of the project. The planning schedule should include the following factors:

1. Total completion time for the project.
2. Earliest time when each activity can start.
3. Latest time when such activity can be started without delaying the total project.
4. Float for each activity, *i.e.*, the amount of time by which the completion of an activity can be delayed without delaying the total project completion.
5. Identification of critical activities and critical path.

Basic Scheduling Computations

We shall use the following notations for basic scheduling computations:

(i, j) = Activity (i, j) with tail event i and head event j

T_E or E_i = Earliest occurrence time of event i

T_L or L_j = Latest allowable occurrence time of event j

D_{ij} = Estimated completion time of activity (i, j)

$(E_S)_{ij}$ = Earliest starting time of activity (i, j)

$(E_F)_{ij}$ = Earliest finish time of activity (i, j)

$(L_S)_{ij}$ = Earliest starting time of activity (i, j)
$(L_S)_{ij}$ = Latest starting time of activity (i, j)
$(L_f)_{ij}$ = Latest finishing time of activity (i, j)

The basic scheduling computations can be put under the following three groups :

Forward pass computations (for earliest event time) before starting computations, the occurrence time of initial network event is fixed. Then, the forward pass computation fields the earliest start and earliest finish time for each activity (i, j), and indirectly the earliest expected occurrence time for each event. This is mainly done in two steps.

Step 1. The computations begin from the 'start' node and move towards the 'end' node for carriers the forward pass computations start by assuming the earliest occurrence time of zero for the initial project event.

Step 2.

(*i*) Earliest starting time of activity (i, j) is the earliest event time of the tail end event *i.e.*, $(E_S)_{ij} = E_j$ is the earliest starting time of the activity time *i.e.*, $(E_F)_{ij} = (E_S)_{ij} + D_{ij}$ or $(E_f)_{ij} = E_i + D_{ij}$.

(*ii*) Earliest event time *j* is the maximum of the earliest finish times of all activities ending into that event that is $E_j = \max_i$ [$(E_F)_{ij}$ for all immediate predecessor of (i, j)] or $E_j = \max_i [E_i + D_{ij}]$ the computed 'E' values are put over the respective circles representing each event.

Backward Pass Computations (For Latest Allowable Time)

The latest event times (L) indicates the time by which all activities entering into that event must be completed without delaying the completion of the project. There can be computed by reversing the method of calculation used for earliest event times. This is done in the following steps:

Step 1. For ending event assume E = L. Remember that all E's have been computed by forward pass computations.

Step 2. Latest finish time for activity (i, j) is equal to the latest event time of event *j*, *i.e.* $(L_F)_{ij} = L$.

Step 3. Latest starting time for event is the minimum of the latest start time of all activities origin-ating from that event, *i.e.*,

$$L_i = \min_j [(L_S)_{ij} \text{ for all immediate successors of } (i, j)] = \min_j [(L_F)_{ij} - D_{ij}]$$
$$= \min_j [L_j - D_{ij}].$$

Step 4. Latest starting time of activity (i, j) = the latest completion time of (i, j) – the activity time

or $$(L_S)_{ij} = (L_F)_{ij} - D_{ij} \quad \text{or} \quad (L_S)_{ij} = L_j - D_{ij}.$$

The computed 'L' values are put over the respective circles representing each event.

Determination of Floats and Slack Times

When the network diagram is completely drawn, properly labelled, and earliest (E) and latest (L) event times are computed as discussed so far, the next object is to determine the floats and slack times defined as follows:

There are mainly 3 kinds of floats as given below:

1. **Total float.** The amount of time by which the completion of an activity could be delayed beyond the earliest expected completion time without affecting the overall project duration time.

   Mathematically, the total float of an activity $(i - j)$ is the difference between the latest start time and earliest start time of that activity. Hence the total float for an activity $(i - j)$, denoted by $(T_F)_{ij}$, can be calculated by the formula:

   $(T_F)_{ij}$ = (Latest start-Earliest start) for activity $(i - j)$

   or $(T_F)_{ij} = (L_S)_{ij} - (E_S)_{ij}$ or $(T_F)_{ij} = (L_j - D_{ij}) - E_i$

   where E_i, L_j and D_{ij} are defined in sec. (Basic scheduling computations) this is the most important type of float, because of concerning with the overall project duration.

2. **Free float.** The time by which the completion of an activity can be delayed beyond the earliest finish time without affecting the earliest start of a subsequent (succeeding) activity.

   Mathematically, the free floats for activity (i, j), denoted by $(F_f)_{ij}$, can be calculated by the formula:

   $(F_f)_{ij} = (E_j - E_i) - D_{ij}$

   In other words, free float for $(i - j)$ = (Earliest time for event j – Earliest time for event i) – Activity time for (i, j) this float is concerned with the commencement of subsequent activity.

Remarks

1. We know that: $(T_f)_{ij} = (L_j - E_i) - D_{ij}$. But $L_j \geq E_j$ as latest event time is always greater than or equal to the earliest event time. Therefore

$$(T_F)_{ij} \geq (E_j - E_i) - D_{ij} \quad \text{or} \quad (T_F)_{ij} \geq (F_F)_{ij}$$

Hence for all activities free float can takes values from zero upto total float, but it cannot exceed total float.

2. **Free float** is very useful for rescheduling the activities with minimum disruption of earlier plans.
3. **Independent float.** The amount of time by which the start of an activity can be delayed without effecting the earliest start time of any immediately following activities, assuming that the preceding activity has finished at its latest finish time.

   Mathematically, independent float of an activity (i, j) denoted by $(I_f)_{ij}$ can be calculated by the formula:

   $$(I_f)_{ij} = (E_j - L_i) - D_{ij}$$

   The negative independent float is always taken zero. This float is concerned with prior and subsequent activities.

   Remarks:

   (*i*) It can be observed that independent float ≤ free float ≤ total float.

   (*ii*) The concept of floats is useful for the management in representing under utilized resources and flexibility of the schedule and the extents to which the resources will be utilized on different activities.

   (*iii*) The float can be used for re-deployment of resources to level the same or to reduce project duration. However, one should bear in mind that whenever the float in a particular activity is utilized, the float of not only that activity but that of other activities would also change.
4. **Interfering float.** Utilization of float of an activity may affect the float times of the other activity in the network. Interfering float is that part of total float which causes a reduction in the float of successor activities. It is the difference between the latest finish time of activity in question and the earliest starting time of the following activity or zero whichever is larger. It represents the portion of the float of an activity which cannot be consumed without adversely affecting the float of the subsequent activities.
5. **Event slacks.** For any given event, the event slack is defined as the difference between the latest event and earliest event times. Mathematically, for a given activity (i, j),

   Head event slack $= L_j - E_j$,

   Tail event slack $= L_i - E_i$,

   All the floats defined earlier can be represented in terms of head and tail event slacks also.

Total float $= L_j - E_i - D_{ij}$

Tree float $= E_j - E_i - D_{ij} = (L_j - E_i - D_{ij}) - (L_j - E_j)$

$=$ Total float – Head event slack

Independent float $= E_j - L_i - D_{ij} = (E_j - E_i - D_{ij}) - (L_i - E_i)$

$=$ Free float – Tail event slack.

6. Time scale representation of float and slacks. The various floats and slack for an activity $(i - j)$ can be represented on a time scale as shown in Fig. 5.23.

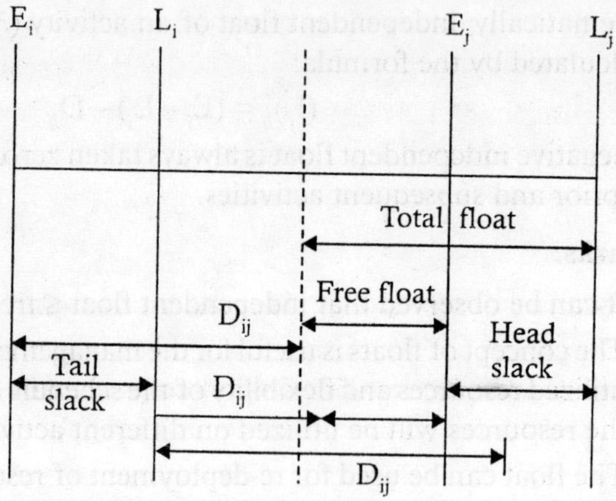

Fig. 5.23

Conclusion draw from total float values:

The value of total float for any activity is useful for drawing the following conclusions:

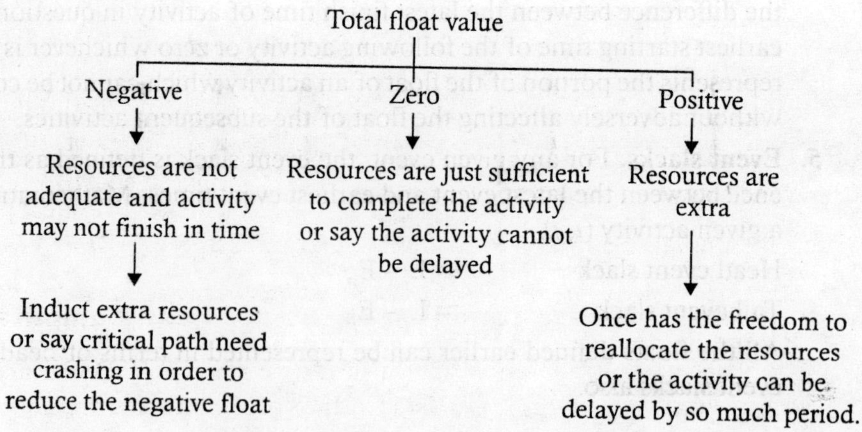

Determination of Critical Path

Before defining critical path, let us first discuss about the meaning of critical event and critical activity.

Critical event. Since the slack of an event is the difference between the latest and earliest event times, *i.e.*, slack (i) = $L_i - E_i$, the event with zero slack times are called critical events.

In other words, the event (i) is said to be critical if $E_i = L_i$.

Critical activity. Since the difference between the latest start time and earliest start time of an activity is usually called as total float, the activities with zero total float are known as critical activities.

In other words, an activity is said to be critical if a delay in its start will cause a further delay in the completion date of the entire project.

Numerical calculation. Consider the network (Fig. 5.20) by assumption

$$E_1 = 0 \quad \text{and} \quad E_2 = -\max_i \,[E_i + D_{j2}].$$

For node 2, node 1 is the only predecessor and hence $j = 1$ contains only one element. Therefore $E_2 = E_1 + D_{12} = 0 + 2 = 2$.

Likewise, values of E_3, E_4, E_5 and E_6 can be computed as:

Consider node 7, where there are 3 emerging activities,

i.e. $$E_7 = \max_i \,(E_i + D_{ij}),$$

The collection i consists of nodes 3, 4 and 6 that are preceding node 7.

Therefore

$$E_7 = \max\,[E_3 + D_{37} = 2 + 8 = 10,\ E_4 + D_{47} = 2 + 4 = 6,$$
$$E_6 + D_{67} = 7 + 0 = 7] = 10$$
$$E_8 = \max\,[E_5 + D_{58} = 6 + 2 = 8,\ E_6 + D_{68} = 7 + 4 = 11] = 11$$
$$E_9 = \max\,[E_8 + D_{89} = 11 + 3 = 14,\ E_7 + D_{79} = 10 + 5 = 15^*] = 15$$

and $$E_{10} = E_9 + D_{9,10} = 15 + 4 = 19.$$

From this computation, it can be inferred that this job will take 19 days to finish as this is the longest path of the network activities along this longest path are: 1 – 3, 3 – 7, 7 – 9 and 9 – 10. This longest path is called the critical path in any network, it is not possible that there can be only one critical path. For example, if in the above network, let D_{36} = 6 days, then two critical paths exist having the same duration for completion of project.

Determination of Latest Time (L_i): Backward Pass Computation

In forward pass computation, the earliest time when a particular activity will be completed is known, it is also seen the some activities are not critical to the completion of the job. The question a manager would like to ask is: can their starting time be delayed. So that the total completion time is still the same? Such a question may arise while scheduling the resources such as: manpower, equipment, finance and so, on. If delay is allowable that what can be the maximum delay? For this is the latest time for various activities which is desired. The backward pass computation procedure is used to calculate the latest time for various activities in the forward to assign the project term in event the date on which the project should be over. If no such date is prescribed, backward pass computation are as follows:

Rule 1. Set $L_i = E_i$ or T_S where T_S is the scheduled date for completion and E_1 is the earliest time.

Rule 2. Repear rule 2 until starting activity is reached.

Rule 3. $L_1 = \min_j \{L_j - D_{ij}\}$, *i.e.* the latest time for activities is the minimum of the latest time of all succeeding activities reducing their activity is reached.

$$L_9 = \min_j \{L_j - D_{9,j}\} = 19 - 4 = 15 \text{ for } j = 10. \quad (j \text{ contains only one node 9})$$

$$L_8 = \min_j \{L_j - D_{8,j}\} = L_9 - D_{8,9} = 15 - 3 = 12$$

$$L_7 = \min_j \{L_j - D_{7,j}\} = L_9 - D_{7,9} = 15 - 5 = 10 \qquad (j \text{ contains node 9}).$$

Now consider node 6. For this node, there are two succeeding activities, namely 6 – 8 and 6 – 7.

$$\text{Hence,} \quad L_6 = \min [L_j - D_{6,j}] = \min \begin{bmatrix} L_7 - D_{6-7} \\ L_8 - D_{6-8} \end{bmatrix} = \min \begin{bmatrix} 12 - 4 = 8^* \\ 10 - 0 = 10 \end{bmatrix} = 8$$

Obviously, a non-critical activity is such that the time between its earliest start and its latest completion dates (as allowed by the project) is longer than its actual duration. In this case, non-critical activity is said to have a slack or float time.

Critical path. The sequence of critical activities in a network called the critical path.

The critical path is the longest path in the network from the starting event to ending event and defines the minimum time required to complete the project.

By the term 'path' we mean a sequence of activities such that it begin at the starting event and end at the final event. The length of a path is the sum of the individual times of the activities lying on the path.

If the activities on critical path are delayed by a day by different means. The critical path is denoted by double or darker lines to make distinction from the other non-critical paths.

Main features of critical path. The critical path has two main features.

(*i*) If the project has to be shortened, then some of the activities on that path must also be shortened. The application of additional resources on other activities will not give the desired result unless that critical path is shortened first.

(*ii*) The variation in actual performance from the expected activity duration time will be completely reflected in one-to-one fashion in the anticipated completion of the whole project. The critical path identifies all critical activities of the project. The method of determining such a path is explained by the following numerical example.

Example 1. Consider the following network where nodes have been numbered according to the Fulkerson's rule numbers along various activities represent the normal time (D_{ij}) required to finish that activity, *e.g.* activity (3) – (6) will take 5 days (months, weeks, hours depending on the time units). For this project, we are interested to find out the time it will take to complete this project.

What jobs are critical to the completion of the project in time, etc.?

Solution. For this it is necessary to find out the earliest and the latest completion time for each activity in the network. The earliest and the latest times are re-calculated by using 'forward pass' and 'backward pass' computations, respectively.

To understand the procedure, we define:

E_i = the earliest expected occurrence time of event i,

L_j = the latest allowable event occurrence time for event j.

This is the latest time by which the event j must be started without increasing the project duration.

D_{ij} = the expected duration to complete the activity $i - j$.

The solution now starts by the forward pass computation.

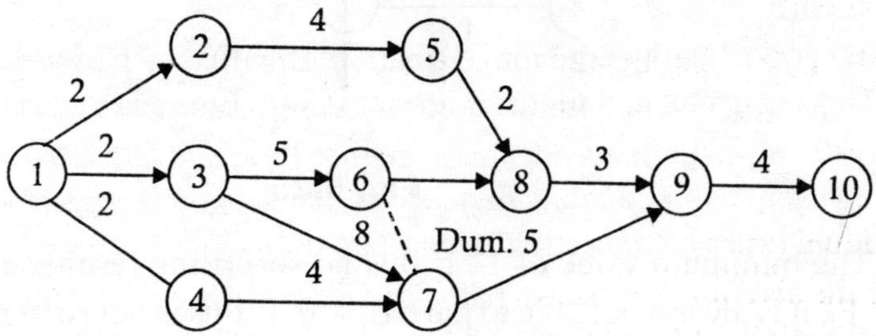

Fig. 5.24

Step 1. Determination of Earliest Time (E_j): Forward pass computation. The purpose of the forward pass computation is to find out earliest start time for all the activities. For this, it is necessary to assign some initial value to the starting node 1. Usually this value is taken to be zero so that the subsequent earliest time could be interpreted as the project duration up to that point in question.

Rules for the computation are as follows:

Rule 1. Initial event is supposed to occur at time equal to zero that is, $E_1 = 0$.

Rule 2. Any activity can start immediately when all preceding activities are completed the earliest time E_j for node j is given by $E_j = \max [E_i + D_{ij}]$, where i is the collection of node which precede node j.

Rule 3. Repeat step 2 for the next eligible activity until the end node is reached.

Similarly $L_5 = L_8 - D_{5,6} = 12 - 2 = 10$, $L_4 = L_7 - D_{4,7} = 10 - 4 = 6$

$$L_3 = \min \begin{bmatrix} L_6 - D_{3,6} \\ L_7 - d_{3,7} \end{bmatrix} = \min \begin{bmatrix} 8-5=3 \\ 10-8=2* \end{bmatrix} = 2,$$

$$L_2 = L_5 - D_{2,5} = 10 - 4 = 6$$

$$L_1 = \min_{j=(2,3,4)} [L_j - D_{1,j}] = \min \begin{bmatrix} 7-5=5 \\ 2-2=0* \\ 6-2=4 \end{bmatrix} = 0.$$

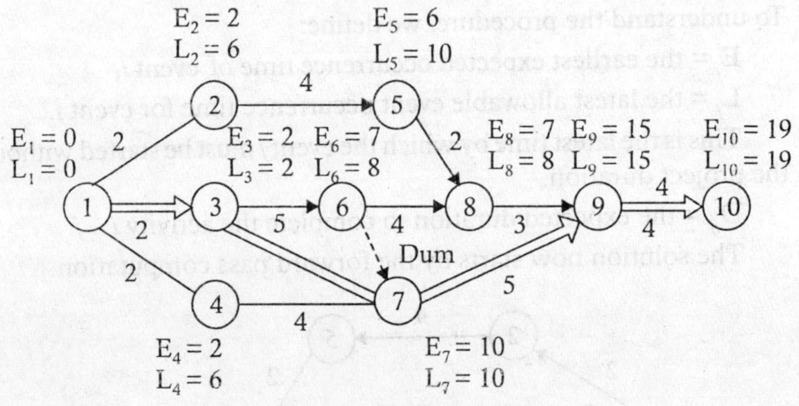

Fig. 5.25

The minimum value of $L_1 = 0$ is no surprising result, since, started with $L_i = E_i$, it is always possible to have $L_1 = 0$. If this is not so, it means that some error is made in calculations of forward pass or backward pass values. (Fig. 5.25) shows earliest and latest times of each event.

Recall the path 1-3-7-9-10 was defined as the critical path of this network. Along this path, it is observed that the latest and earliest times are the same implying that any activity along this path cannot be delayed without affecting the duration of the project.

Step 2. Computation of floats. By definition for the activity 8 – 9, the float is one day ($L_8 - E_8 = 12 - 11 = 1$). This float represents the amount by which this particular activity can be delayed without influencing the duration of the project.

Also, by definition, free float, if any, will exist only on the activities merge points to illustrate the concept or free float, consider path 1-2-5-8-9, total float on activity 8-9 is one day and since this is the last activity prior to merging two activities, this float is free float also. Similarly, consider the activity 5-8 which has a total float of 4 days but has only 3 days of free float because 1 day of free float is due to the activity 8-9. If the activity 5-8 is delayed upto three days, the early start time of no activity in the network will be affected. Therefore, the concept of free float clearly states that the use of free float time will not influence an succeeding activity float time.

If free float of any activity comes out to be negative it is taken zero.

For example, independent float of (1, 2) = free float of (1, 2) – ($L_1 - E_1$) = 0 – (0 – 0) = 0.

Step 3. To identify critical path. The earlier calculation shows that the path or paths which have zero float are called the critical ones. This logic is extended little further, it would provide a guide rule to determine the next most critical path and so on. Such an information will be useful for mangers in the control of projects. In this example, path 1-3-8-9-10 happens to be next to critical; path because it has float of one day one many of its activities.

Activity	Duration	Start		Finish		Float		
(i – j)	(D_{ij})	Earliest	Latest	Earliest	Latest	Total	Free	Independent
(1)	(2)	(3) E_i	(4) = (6) – (2)	(5) = (3) + (2)	(6) L_j	(7) = (4) – (3)	(8) = $E_j - E_i - D_{ij}$	(9) = (8) – ($L_i - E_i$)
1-2	2	0	4	2	6	4	0	0
1-3	2	0	0	2	2	0	0	0
1-4	2	0	4	2	6	4	0	0
2-5	4	2	6	6	10	4	0	0
3-6	5	2	3	7	8	1	0	0

3-7	8	2	2	10	10	0	0	0
4-7	4	2	6	6	10	4	4	0
5-8	2	6	10	8	12	4	3	0
6-8	4	7	8	11	12	1	0	0
7-9	5	10	10	15	15	0	0	0
8-9	3	11	12	14	15	1	1	0
9-10	4	15	15	19	19	0	0	0

The method discussed earlier adopatable on computer. In the case of small network, we can perform most of the calculation right on the diagram in an event that a person would like to use table all format to find floats etc., such methods are also available. Summarises float times and other information.

Example 2. A project consists of a series or tasks labelled A, B, ..., H, 1 with the following relationships (W < X, Y, means X and Y cannot start until W is completed; X, Y < W means W cannot start until both X and Y are completed) with this notation, construct the network diagram having the following constraints : A < D, E; B, D < F; C < G; B < H; F, G < I.

Find also the minimum time of completion of the project, when the time (in days) of completion of each task is as follows:

Task :	A	B	C	D	E	F	G	H	I
Time :	23	8	20	16	24	18	19	4	10

Solution. The resulting network is shown in (Fig. 5.22). The dummy activities D_1 and D_2 are introduced to established the correct precedence relationship the nodes of the project are numbered so that their ascending order shows the direction of progress in the project:

To determine the critical path (minimum time of completion of the project), compute the earliest start E_i and latest finish L_j for each task (i, j). The calculations are done as follows:

$$E_1 = 0, E_2 = E_1 + D_{12} = 0 + 20 = 20, E_3 = E_1 + D_{13} = 23 + 23$$

To determine the value of E_4, since there are two incoming tasks (1, 4) and (3, 4),

$$E_4 = \max_{j=1,3} [E_i + D_{i4}] = \max [E_1 + D_{14}, E_3 + D_{34}] = \max$$

$$[0 + 8, 23 + 16] = 39$$

This procedure is repeated to compute all E_j. Thus

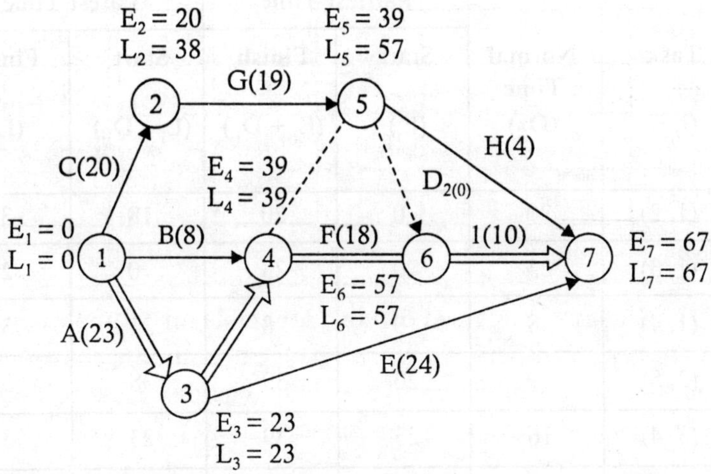

Fig. 5.26

$$E_5 = \max_{i=2,4} [E_i + D_{15}] = \max [39 + 0, 20 + 19] = 39,$$

$$E_6 = \max_{j=4,5} [E_i + D_{i6}] = \max [39 + 18, 39 + 0] = 57$$

$$E_7 = \max_{i=3,5,6} [E_i + D_{i7}] = \max [23 + 24, 39 + 4, 57 + 10] = 67.$$

The value of L_i are calculated proceeding backwards as follows:

$$L_7 = E_7 = 67, L_6 = L_7 - D_{67} = 67 - 10 = 57$$

$$L_5 = \min_{j=6,7} [L_j - D_{5j}] = \min 57 - 0, 67 - 4 = 57$$

$$L_4 = \min_{j=5,6} [L_j - d_{3j}] = \min [57 -0, 57 - 18] = 39$$

$$L_3 = \min_{j=4,7} [L_j - d_{3j}] = \min [39 - 16, 67 - 24] = 23$$

$$L_2 = L_5 - D_{25} = 57 - 19 = 38$$

$$L_1 = \min_{j=2,3,4} [L_j - D_{1j}] = \min [38 - 20, 23 - 23, 29 - 8] = 0$$

Table 5.2. Network Analysis Table

Task	Normal Time	Earliest Time		Latest Time		Float Time
		Start	Finish	Start	Finish	
(i, j)	(D_{ij})	(E_i)	($E_i + D_{ij}$)	($L_i - D_{ij}$)	(L_j)	($L_j - D_{ij}$) $- E_i$
(1, 2)	20	0	20	18	38	18
(1, 3)	23	0	23	0	23	0
(1, 4)	8	0	8	31	39	31
(2, 5)	19	20	39	38	57	18
(3, 4)	16	23	39	23	39	0
(3, 7)	24	23	47	43	67	20
(4, 5)	0	39	39	57	57	18
(4, 6)	18	39	57	39	57	0
(5, 6)	0	39	39	57	57	18
(5, 7)	4	39	43	63	67	24
(6, 7)	10	57	67	57	67	0

From this table critical nodes are for the tasks (1, 3) → (3, 4) → (4, 6) → (6, 7). Thus critical path is 1 → 3 → 4 → 6 → 7 as shown by double line arrow ( ) in the figure. This path represents the longest possible duration to complete the entire project.

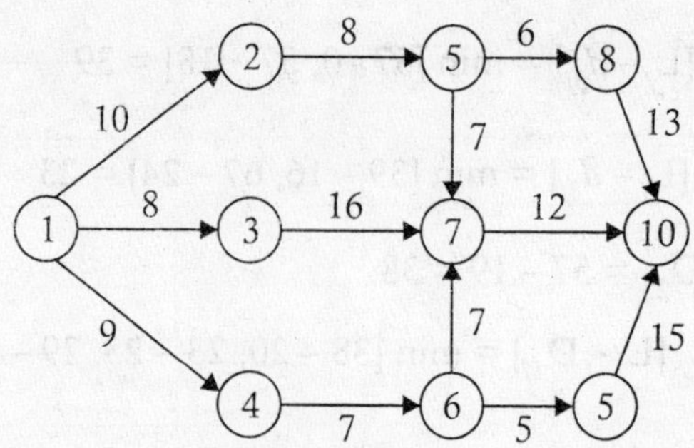

Fig. 5.27

Example 3. Find the critical path and calculate the stack time for each event for the following PERT diagram.

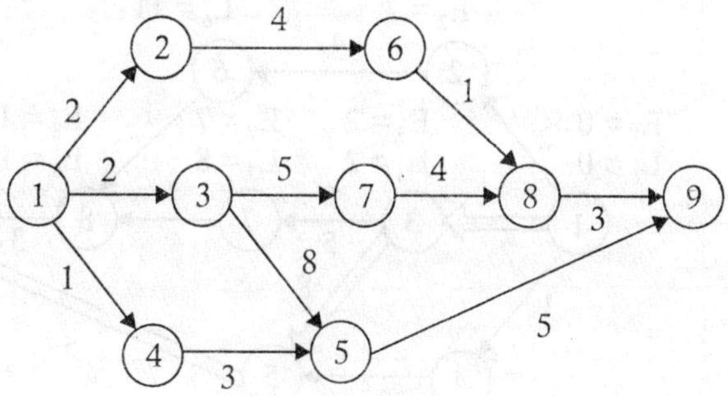

Fig. 5.28

Solution. Proceeding as in example 3, the earliest times and latest times in respect of node points are obtained below:

Table 5.3. Network Analysis Table

Activity (i, j)	Normal Time (D_{ij})	Earliest Time Start (E_i)	Earliest Time Finish ($E_i + D_{ij}$)	Latest Time Start ($L_j - D_{ij}$)	Latest Time Finish (L_j)	Float Available ($L_j - D_{ij}$) $- E_i$
(1, 2)	2	0	2	5	7	5
(1, 3)	2	0	2	0	2	0
(1, 4)	1	0	1	6	7	6
(2, 6)	4	2	6	7	11	5
(3, 7)	5	2	7	3	8	1
(3, 5)	8	2	10	2	10	0
(4, 5)	3	1	4	7	10	6
(5, 9)	5	10	15	10	15	0
(6, 8)	1	6	7	11	12	5
(7, 8)	4	7	11	8	12	1
(8, 9)	3	11	14	12	15	1

From above table critical nodes are the activities (1, 3) (3, 5) and (5, 9).

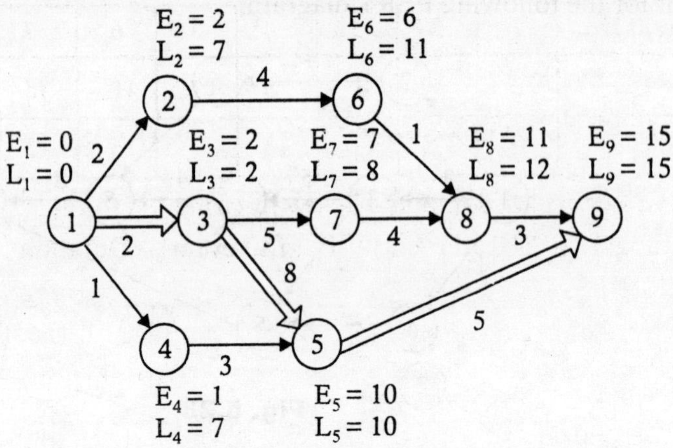

Fig. 5.29

The critical path will be $1 \rightarrow 3 \rightarrow 5 \rightarrow 9$.

Example 4. A project has the following times schedule:

Activity	Time in Weeks	Activity	Time in Weeks
1-2	4	5-7	8
1-3	1	6-8	1
2-4	1	7-8	2
3-4	1	8-9	1
3-5	6	8-10	8
4-9	5	9-10	7
5-6	4		

Construct PERT network and compute:

(*i*) T_E and T_L for each event

(*ii*) Float for each activity

(*iii*) Critical path and its duration.

Solution. The network is constructed as given in (Fig. 5.26).

The T_E's and T_L's computed on the network are as follows:

Event No.	1	2	3	4	5	6	7	8	9	10
T_E	0	4	1	5	7	11	15	17	18	25
T_L	0	12	1	13	7	16	15	17	18	25

Activity floats are computed below by using the formula:

Float = T_L (Head event) – T_E (Tail event) – Duration

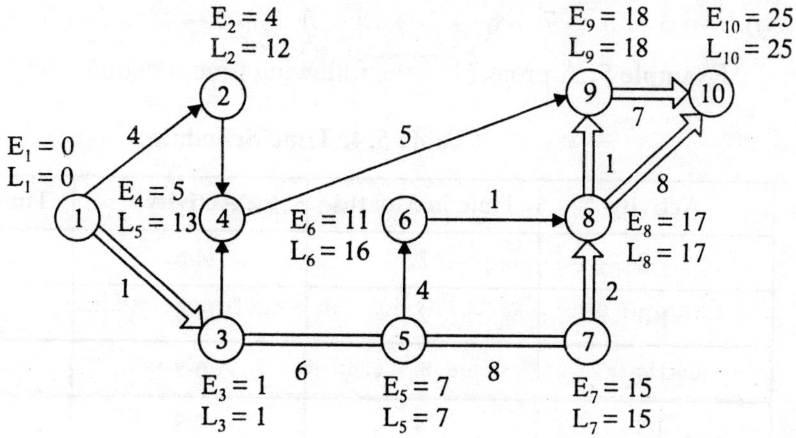

Fig. 5.30

Activity	Duration	T_E (Tail Event)	T_L (Head Event)	Float
1-2	4	0	12	8
1-3	1	0	1	0
2-4	1	4	13	8
3-4	1	1	13	11
3-5	6	1	7	0
4-9	5	5	18	8
5-6	4	7	16	5
5-7	8	7	15	0
6-8	1	11	17	5

7-8	2	15	17	0
8-9	1	17	18	0
8-10	8	17	25	0
9-10	7	18	25	0

Along the zero-float activities there are two critical paths as indicated by double line arrows in the network.

(*i*) $1 \to 3 \to 5 \to 7 \to 8 \to 9 \to 10$ (*ii*) $1 \to 3 \to 5 \to 7 \to 8 \to 10$.

Example 5. A project has the following time schedule.

Table 5.4. Time Schedule

Activity	Time in Months	Activity	Time in Months
1-2	2	4-6	3
1-3	2	5-8	1
1-4	1	6-9	5
2-5	4	7-8	4
3-6	8	8-9	3
3-7	5	—	—

Construct PERT network and compute:

(*i*) Total float for each activity (*ii*) Critical path and its duration.

Also find the minimum number of cranes the project must have for its activities 2-5, 3-7 and 8-9 without delaying the project then, is there any change required in PERT network? If so indicate the name.

Solution. The network is constructed as below:

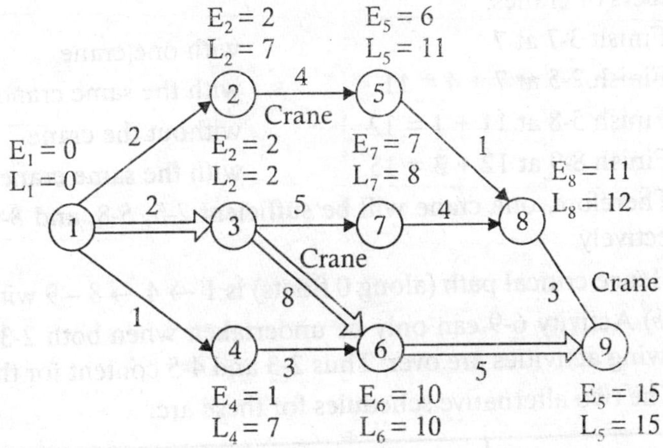

Fig. 5.31

The network analysis table is computed below:

Activity	Duration	Earliest Time		Latest Time		Total Float
	(D_{ij})	Start (E_i)	Finish $(E_i + D_{ij})$	Start $(L_i - D_{ij})$	Finish (L_j)	$(L_j - D_{ij}) - E_i$
1-2	2	0	2	4	6	4
1-3	2	0	2	0	2	0
1-4	1	0	1	6	7	6
2-5	4	2	6	7	11	5
3-6	8	2	10	2	10	0
3-7	5	2	7	3	8	1
4-6	3	1	4	7	10	6
5-8	1	6	7	11	12	5
6-9	5	10	15	10	15	0
7-8	4	7	11	8	12	1
8-9	3	11	14	12	15	1

Thus the critical path is $1 \rightarrow 3 \rightarrow 6 \rightarrow 9$ with duration 15 months. Minimum numbers of cranes:

Finish 3-7 at 7 with one crane

Finish 2-5 at $7 + 4 = 11$ with the same crane

Finish 5-8 at $11 + 1 = 12$ without the crane

Finish 8-9 at $12 + 3 = 15$ with the same crane

Therefore, one crane will be sufficient 2-5, 5-8, and 8-9 start at 7, 11, 12 respectively.

Hence critical path (along 0 floats) is $1 \rightarrow 4 \rightarrow 8 - 9$ with duration of 15.

(*b*) Activity 6-9 can only be undertaken when both 2-3 and 4-5 and their following activities are over. Thus 2-3 and 4-5 content for the equipment.

The two alternative schedules for these are:

Activity	Start	Activity	Start
2-3	2	4-5	2
4-5	6	2-3	7
6-9	15	6-9	12

The second alternative does not delay the project completion time and hence to be recommended.

Examples on Optimum Duration and Minimum Duration Cost

Example 6. Table below shows; jobs, their normal time and cost, and crash time and cost for a project.

Job	Normal Time (days)	Cost (Rs.)	Crash Time (days)	Crash Cost (Rs.)
1-2	6	1400	4	1900
1-3	8	2000	5	2800
2-3	4	1100	2	1500
2-4	3	800	2	1400
3-4	dummy	—	—	—
3-5	6	900	3	1600
4-6	10	2500	6	3500
5-6	3	500	2	800

Indirect cost for the project is Rs. 300 per day.

(*i*) Draw the network of the project.

(*ii*) What is the normal duration cost of the project?

(*iii*) If all activities are crashed, what will be the project duration and corresponding cost?

(*iv*) Find the optimum duration and minimum project cost.

Solution. (*i*) Network is shown below.

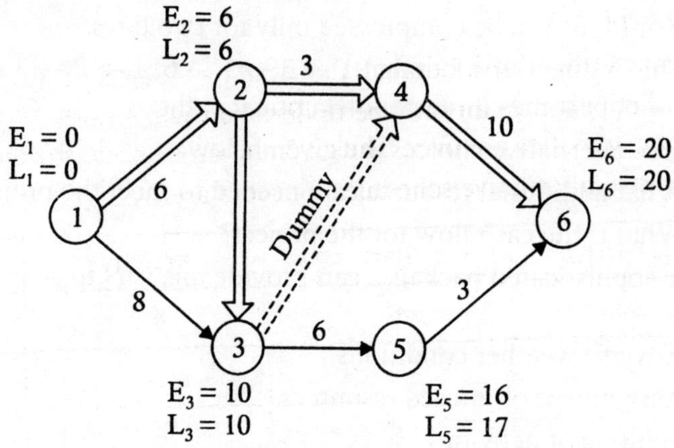

Fig. 5.32

(*ii*) Assuming that all activities occur at normal times, the critical path calculations are shown in the figure under normal conditions. The critical path is $1 \rightarrow 2 \rightarrow 3 \rightarrow 4 \rightarrow 6$ the duration of the project is 20 days and its associated (normal) cost is Rs. 9200.

(*iii*) Now compute the different minimum cost schedule that can occur between normal and crash times mainly depending on the cost time slopes for the different activities to calculate these use the formula.

$$\text{Cost slope} = \frac{\text{crash cost-normal cost}}{\text{normal time-crash time}}$$

These slopes are summarised in the following tabular form:

Activity:	(1, 2)	(1, 3)	(2, 3)	(2, 4)	(3, 5)	(4, 6)	(5, 6)
Slope:	250	267	200	600	233	250	300

Step 1. Because the present schedule involves more time, the schedule is reduced by crashing some of the activities. As the activities lying on the critical path control the duration of the project, therefore the duration of some activities lying on the critical path is reduced.

Start reducing the duration of that activity which involves minimum cost slope as the activity (2, 3) has the minimum cost slope, the duration of this activity is reduced from 4 to 2 days resulting an additional cost of Rs. 2 × 200 = Rs. 400. But this activity should be shortened only by one day, since the path 1 → 2 → 4 → 6 becomes a parallel critical path so the revised schedule corresponds to 19 days with a cost of Rs. (9200 + 200) = Rs. 9400.

Step 2. Now it is evident that the activities (1, 2) and (4, 6) among the remaining activities lying on the two critical paths have the least slope. Therefore either (1, 2) or (4, 6) can be compressed only for two days.

This is due to the fact that 1 → 3 → 5 → 6, 1 → 2 → 3 → 4 → 6 and 1 → 2 → 4 → 6, becomes three parallel critical paths.

So three alternative choices are given below:

- What additional resources are needed to meet the project constraints?
- What is the cash flow for the project?

Some sophisticated packages can provide answers to schedule and cost based upon:

- Adverse weather conditions.
- Assignment of unused resources.
- Splitting of activities.
- Varying the crew size.

Computerisation has also increased the speed at which networks can be formed. Thus the formation and manipulation of the networks can be done without involving too much of manpower as the changes can be observed on a real time basis on the computer.

Another major advantage of having computerised project management is with regards to the complex calculations that need to be performed. A large amount of financial calculations need to be done to impress the financier to fund the project and a lot of projected figures need to be shown. These calculations cannot only be done quickly but also accurately with the aid of computers. The graphs and charts can also be plotted very easily such that they change. Simultaneously in case of any change in the financial values.

Computer Based Project Management Systems

Table 5.5. Lists the programmes, their capabilities, inputs, outputs and mode of operation to provide an overall view of the so-called scientific system available for project management. The list no doubt is impressive and there cannot be any doubt that which such tools and techniques project management can never remain a trick as illustrated in Fig. 7 and remember we have not listed even one-tenth of the number of systems readily available in the market. Things must indeed be too bright for the future project managers. But one must admit that it is not so, at least not today. We may look forward to tomorrow when computer terminals will be available at the project manager's table like we have telephones today. But then there telephones must work, as they, as they are supposed to, and not behave like the ones we see today in our country science and scientific thinking have progressed quite far but it will take us a long time to catch up with that or at heart.

1. Compress (1, 2) by 2-days at a cost of Rs. 250.
2. Compress (4, 6) by 2-days at a cost of Rs. 250.
3. Compress (1, 2) and (4, 6) at a cost of Rs. 250 each.

   The additional cost thus will be Rs. 2 × 250 = Rs. 500 thus a 17 days least cost schedule is obtained with a cost of Rs. (9400 + 500) = Rs. 9900.
4. To determine the optimum schedule, compute the total cost by adding, the direct cost corresponding to each schedule to the cost of crashing (slope) optimum schedule (duration) is then obtained for which the total cost is required calculations are put in the following tabular form:

Normal Project Length (days)	**Crashing Time and Cost (days/Rs.)**	**Indirect Cost @ Rs. 300**	**Total Cost (Rs.)**
20	—	20 × 300	6000
19	1 × 200 = 200	19 × 300	5900
18	1 × 250 = 250	18 × 300	5650
17	1 × 250 = 250	17 × 300	5350
16	1 × 200 + 1 × 600 + 1 × 233 = 1033	16 × 300	5833

Since minimum total cost is Rs. 5350, the optimum duration of the project is 17 days.

Computerised Project Management

With projects becoming tougher and more complex, making analysis has also become very difficult as the amount of information that has to be understood has also increased. This complexity generated a need for a system that could provide results immediately when a certain data was available such a need was met by the advent of computers and the use of the idea of computerised project management çomputerised project management systems can provide answers to question like:

- How will the project be impacted by limited resources?
- How will the project be impacted by a change in requirements?
- What is the impact of overtime?

Develop the desired scientific temperament. And for those who are superstitious, non-believers, peddlers of crafty management or petty politics, these systems may never become a reality even though computer prices may crash to those of telephones or pocket calculators.

Table 5.5. Computerised project management system

Sl. No.	Name of Programme	Programme Description	Mode	Input	Output
1.	Action list	Designed for use by project managers in expediting, progress chasing and for document control, the package is used primarily to establish schedules itemising information outstanding, drawing requirements or materials procurement programmes. These define what needs to be done, by whom and when. The programme is then updated regularly and used for progress chasing, recording what has been done and highlighting critical items needing action.	Inter active	Data entry through screen—with facilities for retrieving and editing any item of data	A variety of tabular and bar chart reports on both the screen and the printer.

2.	Apecs/ 8000	An integrated, interactive multi-user project management package based on a true relational data base facilitating top-down, bottom-up planning. The system is modular and can include critical path analysis with 1-J or precedence networks, progress updating, resource analysis, resource levelling and cost analysis. It can perform WBS or organisational structure analysis, with implicit/explicit structure and multi-level consolidating performance measurement can be carried out using earned value analysis and variance analysis. It has a common field dictionary the system allows user-defined applications in addition to its integral data management functions.	Inter active	Fixed/free format data entry direct or from file, custom screen formated data entry and modification.	Text/tabular, user-defined standard, screen or hard copy, summarises/consolidations. Terminal graphics: bar chart, graphs, logic diagrams, platter graphics Gantt, linked Gantt, S. Curves, logic.

3.	Arches	Develops a contractor's estimate the package contains preliminary design procedures for the structure, finishes, HVAC, electrical, plumbing and cost bases for materials, manpower and indirects. This allows the development of design quantities, material cost and installation crews for each field task. The package develops the manhours for each field craft for site development, field erection and installation of each design quantity. It also prepares a management-oriented report, with key costs backed up with detailed reports of quantities, summaries of installation tasks, material and field manpower costs.	Inter active and batch	Command driven data input generator available	Master summary site-development report, building summary report, code of accounts field manpower summary manpower and construction schedule, cash flow summary prime contractor's indirect costs, equipments rental materials and field power summary.

4.	Capital projects financial planning system	Designed to control the phasing of projects within cash flow and resource constraints. System allow phased budgets to be compared with actual and latest estimated expenditure, and permits 'what-if' analysis of the effect of adding, omitting or rescheduling projects in the programme. Features include multiple currency and escalation and de-escalation capabilities, user defined cost centres and summary cost centres, multiple user defined accounting periods and extensive facilities for analysing variances and slippages.	Batch and interactive	Screen forms or files	Used defined tabular chart and graphic
5.	CBACS-PERT	Functions are to record, review and then re-estimate the status of each task at each stage of a project. To monitor identify areas where time scales, resources	Interactive	On-screen edit; disc file merging on line project data	A variety of report to a wide range of status and sequencing criteria output to screen, project disc.

		demands of project cost targets are in jeopardy. To permit comprehensive reporting by a variety of user criteria to assist with the project planning and control objective to allow interactive modification of strategies for optimising project progress activity time schedules, resources aggregations and project cost estimates can all be calculated and reported in suitable formats.			
6.	C-Cost	A project control programme that conforms in full to the 45 department of defence cost/schedule control systems criteria. Although, C-Cpec techniques are associated, mainly with large defence contracts, they have also been widely defence contracts, they have also been widely applied to projects of all sizes	Batch and interactive	Screen forms and files.	User defined tabular chart and graphic.

		in many other fields. The package is based on the foundation of C-Specially with a number of additional features including document management through word.			
7.	Concept project management	Processor links, user defined calculations, multiple earned value techniques; multiple currencies rates and inflations profiles comprises project planning, resources management and budgeting programmes, precedence network analysis creates bar charts that can be modified on the computer screen and stored, and from which resource histograms and cost information can be generated. Multiple calendars allow for seasonal variations in productivity. The budgeting programme produces	Real time	Key board entry; editing available at all stages.	Extensive output including user designed formats saved for future use; formats may be tabular or based on bar charts, histograms curves and diary pages; budgets financial reports and summarises conform to normal document layout all output standard printers.

		cost estimates, departmental budgets, current total cost estimates and cash flow information. Detailed costing can be generated from the budgets/based on cost-head allocations. Break even point analysis and cost-to-date analysis are easily produced and can be illustrated graphically.			
8.	CRIT path	Critical path project management package. Analysis and critical path network, calculating early start/finish and later start/finish dates.	Interactive	Data entry and editing on screen or option from system data file.	Screen output (tabular) printer output (tabular): plotter output (graphical Gantt/bar chart; system data file output.
9.	Diagonal network analysis	Draw a critical path network to scale on a plotter using simple data entry. Preliminary drawings and calculations are unnecessary since the programme carries out both automatically once the chart has been created it can be adjusted by the user to suit his	Batch plotting, interactive data entry	Interactive editor	Plotter graphics

		requirements, and if the cost of each activity is entered, the chart will draw a running cost curve. When the project starts, data can be entered to show the difference between planned and actual progress, subsequent activities being adjusted to take account of the difference			
10.	Drawing register	Package maintains a record of all drawings produced for a project. It can produce a complete history of any drawing or group of drawings. This shows the file, size, initials of persons responsible for drawing, checking and approving, revision, date and any notes. The report also shows the dates on which the drawing was issued, the recipients and the quantities. The drawing status report indicates the latest revision and status	Interactive	Data entry through prompted screens with facilities for retrieving and editing any item of data	A variety of tabular reports on both the screen and the printer.

		of each drawing and is the main control document for drawings on the project. The system produces a print delivery note that lists all the drawing required.			
11.	Internet 805	A modular suit of programmes that covers operational planning, resource utilization, procurement performance monitoring, cost analysis and financial forecasting. Included is a system generator that enables new management systems to be created. The package supports precedence and arrow networks and has flexible activity descriptions for complex coding structures resources scheduling facilities skeletorisation and linking are available. Network analysis modelling language for resource or cost calculations.	Batch and interactive	Free format input	Single or multiple project reporting using user specified formats; tabular bar charts, histogram or plotted network forms: output to screen and hard copy.

12.	Manpower expenditure recording	Provides computerised control and analysis of manhours booked to projects and overheads within a company or department. It can be applied in any project or cost centre-oriented organisation. The principal input document is a weekly time-sheet prepared by each person. A range of reports can be generated giving the weekly and cumulative manhour totals by project, overhead, section activity and staff agency in addition, project billing and department costing reports can be produced.	Interactive	Data entry through prompted screens	A variety of tabular reports on both the screen and the printer.
13.	Mentor	Comprises project planning resource management and bill of quantities programmes. Precedence network analysis creates bar chart that can be modified on the computer screen and stored and	Real time	Key board entry or bill of quantities from library options: editing and full word processing where appropriate	Extensive output including user designed formats saved for future use formats may be tabular or based on bar charts, histograms, curves and diary pages: bills of quantities

		from which resource histograms and cost information can be generated. Multiple calendars allow for seasonal variations in productivity the billing module produces interim statement from the measured work, approximate valuations and cash flow information detailed costing can be generated from the bill, based upon a breakdown of the rates. Bills of quantities can be compiled from a library module of phrases or item descriptions, or the contract documents.			financial reports summaries conform to normal documents layout; all output to standard printer.
14.	MISTER	Projects analysed during planning and performance phases potential time, resource and cash flow problems identified before their occurrence as alternative options are investigated. The packages tracks the progress and shows the impact of changes in	Interactive and batch	Interactive and direct input	Printed or plotted graphs; plots; bar charts or histograms; printed or screen reports.

		the plan or irregular events that might upset the plan. Selective reporting capabilities provide information for the use of all management and operating levels. Schedules sorted in any number of distinct criteria system is modular; nine modules available in five model sizes.			
15.	N5500	Features include handling of precedence and I-J networks, upto 40,000 in user defined hierarchy sub-projects, phases unlimited milestone events, complex, network logic and hammocking multiple calendars and user define working patterns. Budgeting, cost planning, control and forecasting automatic resource allocation, load levelling and resource planning progress reporting based on a turn-around documents. Its available	Batch and interactive	Data entry by screen or paper input, on line file screen and updating full validation against master file.	Fixed format; flexible content reports: bar charts, S-curve comprehensive network plotting 14 screen based reports: Simulation, on line interrogation reports upto nine levels of summarisation.

		multilevel reporting, historical information and simulation capabilities are available. The needs of matrix organisations are catered for by means of a data base.			
16.	PERT 6	Projects are defined in a top down approach allowing the user to define the level of detail and the number of levels of details and reporting. Each project may use upto 14 user-defined calendars and unlimited resources. Resources levelling is done on a daily basis or on any quantum of time defined by the user. It can communicate with other data bases on the same or separate hardware.	Real time batch and inter active	Input from screen or file; network can be input graphically.	User defined reports include project level, milestone level, activity level, bar charts, earned value, curves, network plot.
17.	Project manager's work station	Provides local data entry and analysis, either as a freestanding data-manager or linked to a company's central	Interactive and batch	Screen forms or files	User defined reports.

		project management facilities includes micro-to-host communications, software and data base management software that can be tailored to individual applications.			
18.	Projects	A modular package consisting of schedule, cost, graphics and relational data base modules. Scheduling establishes network logic, checks for errors and calculates original and current schedules multiple calendars, target scheduling, micro-scheduling, discontinuous scheduling, automatic network generation and sequence scheduling are available, network scheduling based on 'what-*if*' simulations. The cost processor models a project with multiple level hierarchies. The user can measure project performance forecast variances, highlight trends and calculated	Real time batch and interactive	Menu-driven screen; command language.	60 standard reports, customs report writer screen display of reports : Gantt bar charts/network diagrams, histograms: flow curves.

		earned value. The graphic processor supports the display of multiple calendars, micro scheduled projects and automatically produces calendars appropriate for each network. A password access security system is used.			
19.	The planner	A project planning tool employing the principles of critical path analysis with a number of enhancements and extensions. Full resource levelling is included as standard one feature facilitates the direct construction of a project on screen without recourse to drawing the network diagram. Project models may be plotted graphically providing clear records of the logic structure. Sections of projects may be moveder bloc from project to project allowing the establishment of 'logic libraries' containing frequently used data.	Interactive	Keyboard data entry with full editing facilities of network and resource files.	Bar charts, histograms and various reports are sent to printers: optional plots of networks.

20.	Vision	An information package that includes techniques for integrating cost, resource and schedule information. It includes modules for performance measurement to DOD, DOE and NASA standards, and graphical output in the form of networks, bar charts and management graphs.	Interactive	Interactive or batch; can accept files created by other systems.	Colour graphics output of chart and plots on screens or plotters : full and concise formate reports from sorted and selected data.
21.	VUE	An interactive project management package designed to help managers establish and monitor project-management plans. It uses the critical path method highlighting the most important activities. The project managers can modify information and see the project from several perspectives generate several 'what-if' scenarios and respond to unexpected developments with new schedules and strategies.	Interactive	On screen	Output can be on terminals printer or plotter graphic options available.

22.	ARTEMIS	A fourth generation language with special facilities for project management. The language which is identical for all applications, used English—Like commands provides the facility to formulate adhoc enquiries and can be used as menus for inexperienced users. A relational data-base enables the rapid implementation of application: mirroring existing data structure and providing for subsequent modification without the need for reimplementation it has in built procedures for network planning and resource scheduling, aggregation across time, and data entry checking and validation. A report writer and project graphics facilities are included.	Interactive, batch and real time	User definable formatted screen; unformated fast entry; offline data preparation.	User definable text/tabular reports, bar charts, histograms pi-charts, network plots etc.; colour-graphic reports available as required; output can be to screen, printer or plotter.

Review Questions

1. Discuss how Gautt Chart is used as a Tool in Project Management?
2. With a neat sketch explain the Logic flow for production of a work piece.
3. Give the applications of PERT/CPM techniques.
4. Explain the importance of computers in Project Management.
5. Short notes:
   (*i*) Fulkersons Rule
   (*ii*) Steps involved in PERT/CPM
   (*iii*) Network diagram representation.

6 CO-ORDINATION AND CONTROL

For the successful completion of any project, intervention of different types is a must at different levels. The most common forms of intervention are:

1. Direction.
2. Co-ordination.
3. Control.

At the initial stage of the project, direction, will have to be provided:

Project Direction

Project direction refers to the use of authority to channelise the activities of the project on desired lines. Since plans and schedules prescribe what is desirable for time control of a project, authorising them for implementation would constitute project direction towards time control. When schedules are so authorised, they become working documents and depending on the importance attached to them, can even become work orders. Executing agencies or individuals will then be compelled to strictly follow the plans and schedules. If the authorisation is not there, plans and schedules may not go into work.

More development of plans, systems and procedures would not produce better results unless they are authorised for implementation. Any change or deviation from what has been authorised should require approval, as this alone can ensure that the project is proceeding along desired lines. These authorisations and approvals from the key constituents of project direction for the successful implementation of any project. During the initiation or start up period of the project, direction shall be provided by the project manager. Production period directions will be exercised by other members of the project organisation, or else the project will stall. Authority will have to be delegated for this purpose.

Direction During Project Initiation

Project direction is maximum at the time of start up of implementation. This period starts soon after the issue of the project charter. The project charter defines the broad scope of work and overall time and cost target, but does not go into details. The charter authorises the project managers to spell out the details and to issue directives for realisation of the same. Everything at the time of start up is therefore hazy and hence direction is required from the project manager.

The directions given by the project manager at the start up period relates to :

1. Scope of work.
2. Specifications of results of completed work.
3. Schedule of work.
4. Budget for work.
5. Division of work.
6. Systems and procedures for work.
7. Authority and accountability for work.
8. Control and co-ordination of work.

Direction During Production Stage

Direction during the initiation stage establishes the base lines for the project performance. Direction after the initiation period can be considered to be of administrative variety. After the start up period, direction is provided on a case basis through formal documents or personal contacts. A group meeting may also be used for this purpose. Directions may refer to approval of work schedules, detailed budgets, specifications, purchase orders, work order etc. Direction not affecting baselines of project goals can be considered as lower level project directions, it is necessary that lower level directions are provided at appropriate levels down the hierarchy on a day to day basis through personal contact. For matters concerning multiple agencies meetings may be preferred for issuing directions. Under no circumstances should directive authority be too centralised or kept remote from the users.

Direction During Closing Period

By the closing period, the project would have reached the end of its journey and no specific direction from the management would be required.

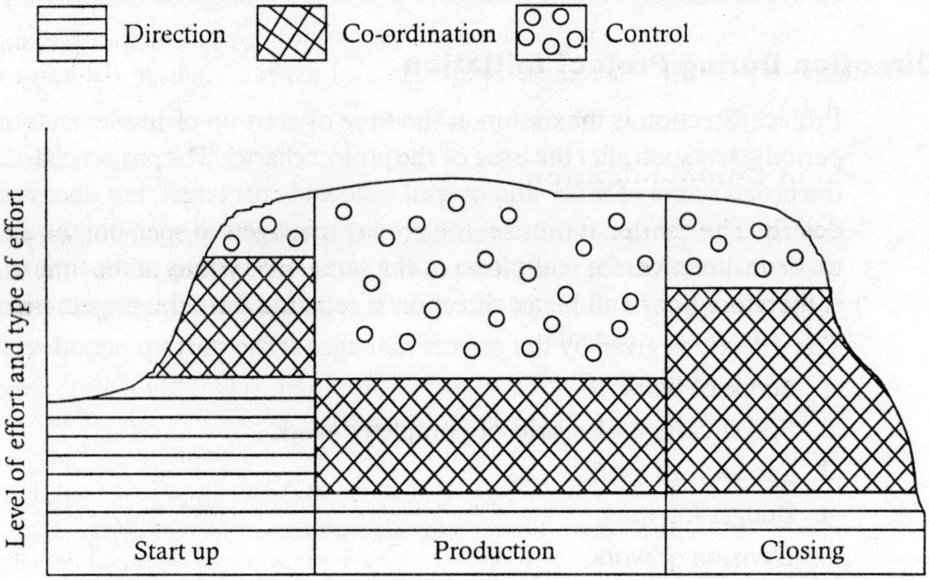

Fig. 6.1

Project Communication

The entire process of direction, co-ordination and control in a project revolves around communication. Through direction weads to co-ordination and control, direction itself is effective only when there are two way communications all around about 60% of management problems are caused in whole or in part by faulty management communication.

Communication does not merely mean passing on a message. But it most include the understanding of the message. If the message sent by the sender is not understood by the receiver, then no communication is said to have taken place. Merely providing the channels for communication is not enough. Effort must be directed towards clarifying and making sure that the receiver has received and clearly understood what was intended to be conveyed.

E.g. A drawing is sent by the sender, but he does not care to check if it has been received and can be put to use. The sender may even send the wrong drawing. The receiver of the drawing also may not acknowledge the receipt of the

Americans have an interesting way of communicating. When one person ask another what he thinks about an issue, the respondent would invariably repeat the question first before making any reply. This is to make sure that he has correctly understood the question.

Communication has two dimensions, physical aspects include passing on a meno, data, instruction, etc. while mental aspects include listening skills, empathy, etc.

Problems in Communication

Since communication is the key to successful project management, every effort must be made to ensure complete communication between all the members of the project. Both the physical and mental aspects of communication must be taken care of some problems are:

1. **Organisation Chart.** The organisation chart is supposed to provide the official communication channel but some organisation charts that are not very clear become a major hindrance for communication.

   Remedy. Departmentalisation and chain of command must be allowed but they should not act as a barrier for communication. The orders must flow along the chain of command but the information flow-must not have any such restrictions.

2. **Physical Distance.** The physical distance between 2 interacting groups will usually present a barrier in communication. Through the people may be working in the same office, shuttling from one cabin to the other can create a problem. The physical distance can also be created by making one self unapproachable even in open work environments.

   Remedy. Have all the working groups in the same office and provide them with good communication equipment.

3. **Absence of Communication Devices.** Sometimes the 2 groups that are working may be such that one is on the site while the other is in the office thus it is necessary to provide both the groups with appropriate communication devices to bridge the distance between the 2 groups.

   Sometimes it is better to over-communicate as to prevent rumours filling the voids created due to the communication voids. Lack of clear communication can also lead to misunderstanding, lack of trust, loss of confidence, etc.

Some actions that can be taken to improve communication are:

1. Selection and installation of appropriate communication devices.
2. Organising review meeting at predetermined frequencies.

3. Installing a structure reporting program.
4. Running an in house magazine.
5. Using desk top computers for communication.
6. Implementing a routine communication system and procedure.

Basic Requirements for a Project Management Information System:

1. Each project header must be able to update his project on a monthly basis utilising project management forms.
2. The project manager must be able to manually device the information needed to measure the progress of the project from the reporting systems.
3. The project leader must be allowed to review his project plans and predict the cash required every week or month to complete the project.
4. Analysis of the reasons in the exception report must be done by the project manager.
5. The analysis reports must be then typed up for distribution.

Optimum Project Management Information System may be described as follows:

- The existing information system are reduced to three elements:
  - Human resources.
  - Financial resources.
  - Material resources.
- Direct entry of data is via terminals with CRT's (cathode ray tube).
- Individual reports may be generated for other than project management needs.
- Information is formulated along with estimates supplied by the project leader to display a current summary, plan and analysis on the CRT or as a hard copy.
- Additional reports could be generated from this information such as facilities scheduling, work planning, manufacturing engineering, etc.
- Project plans can be transferred by project leaders to line management via a CRT.
- The system will have the capability to access central pertinent files.
- The system will have the capability for data correction at the original source, thus reducing the labour required.

Performance Control

A project on its completion must perform. Failure of the project to come up in time or budget can be tolerated, but failure to perform will never be condoned. Personnel are criticised not only for the time and cost over-run, but for non performance they are cursed day in and day out by the plant operation personnel.

Performance control like progress control will require continuous monitoring during the production phase. This must however, be preceded by right specification, selection of right vendors, contractors and above all right contractual stipulations regarding warranties and guarantees. The warranty in this context specifies the quality or performance the purchaser has a right to rely upon. The guarantee concerns what will be done by the seller, contractor or manufacture in the event if goods are found defective or do not meet certain standards of performance. The warranty normally deals with a specific period of time and performance and this is what the purchaser bargains and pays for. The guarantee is a liquidated damage clause that refers to broken warranties. A guarantee costs the seller or contractor or manufacturer money or its equivalent in material parts or services. But, a project cannot simply depend on the suppliers warranties and guarantees. Due to project slippage, the warranty period may often expire before the project is completed. What the owner may get in terms of liquidated damages even if they could ever be realised, would be a poor compensation for the perpetual headache caused by a non performing plant. On going performance control is therefore essential.

Performance control starts with identification of performance parameters. Which performance parameters are critical for viability of the project must be first establish in clear terms. Output certainly is one performance parameter: raw materials and power-consumption per unit of production could be others. The design and specification must be made to meet these performance requirements. The subsequent stages of project execution, like ordering and manufacture must ensure that these requirements are not traded off for some other short term benefits. Thus a single performance parameters would require to be tracked through all the stages of the project execution till the final performance guarantee confirms achievement of the performance.

Schedule Control

Schedule control is to ensure adherence to the agreed time schedule for the project. It is not an independent activity. Schedule has to be used as a basis for direction, communication, co-ordination and progress control or else there can be no schedule control. Schedules and scheduling systems have to form an inseparable part of the project execution for exercise of schedule control.

Why schedule control?

The thrust is to convert all available inputs to outputs by taking commitments from the working groups. But since input for one group has to come from the output of another, all the groups have to progress in a desired way or else, the progress will be uneven. If this happens the project completion will be delayed. It is also possible that a working group has achieved 99% progress and yet due to lack of a single input completion, it can be delayed by several months. Monitoring and control of project and time becomes essential besides progress control to ensure adherence to project schedule.

Types of Schedule Control

1. Preventive schedule control.
2. Predictive schedule control.

Preventive Schedule Control. It is good practice to keep a close watch on the areas which normally delay a project and take all possible steps to make sure that delays do not occur. This means that, project management can use past experiences of similar nature to forecast vulnerable areas and act in advance instead of reacting to the forecast of completion dates of the current project. Delays occur more in the beginnings of the project than towards the end. The delays that one often experience at the later part of the project are usually a hangover of the past. Thus schedule control must start at the beginning of the work and it is here that past experiences will prove invaluable.

Predictive Schedule Control. Starts with the forecast of project completion if the forecast indicates that the project completion requirements are going to be met, no action is initiated. This forecast is to be provided to the controlling authority in the form of the feedback report. If the forecast indicates that the project completion data may not be met, immediate investigations are ordered to find out the ways and means of bringing the project back to schedule.

Cost Control

Cost control is not only monitoring of costs and recording massive quantities of data, but also analysing the data in order to take corrective action before it is too late. Cost control must be performed by all personnel who incur costs. It is the cost which reflects performance in all areas. A well managed project will incur lower cost than one which is not. If the project cost can be brought down without compromising on the parameters that would make a project perform. It would indicate excellent cost management. Cost control therefore should not be mere accounting of costs to report expenditures but must also include monitoring all the cost components thus ensuring the completion of the project at optimal cost.

On going cost reduction must be involved with lowering costs. Without lowering the performance. Any revision in the cost estimate made must be treated as only a revised forecast of the total project cost and must not be treated as licenses for additional expenditure. Cost control should therefore refer to the manipulative efforts during the various phases of the project for reducing cost not yet incurred.

Cost control must include:

- Cost estimation.
- Company cash flow.
- Cost accounting.
- Direct labour costing.
- Project cash flow.
- Overhead rate costing.
- Incentives, penalties and profit sharing.

Project Control

No sooner is the project launched, control becomes the dominant concern of the project manager. Indeed, once the launch phase is over, planning and control become closely interwined in an integrated managerial process.

Project control involves a regular comparison of performance against targets, a search for the causes of deviation and a commitment to check adverse variances. It serves two major functions:

1. It ensures regular monitoring of performance and
2. It motivates project personnel to strive for achieving project objectives.

Reasons for Ineffective Control

Effective control is critical for the realization of project objectives. Yet, control of projects in practice tends to be ineffective. Why? There seems to be three reasons for poor control of projects.

1. **Characteristics of Project.** Most of the projects are large, complex undertaking involving many organisations and people. This renders the task of control difficult because:
   (*a*) Keeping track of physical performance and expenditure on hundred or thousand (or of thousands) of activities which are often non-routine is a stupendous task.
   (*b*) Co-ordinations and communication problems multiply when several organisations are involved in the project.

2. **People Problems.** To control a non-routine project, a manager requires an ability to monitor a wide range of disparate factors, a sensitivity to symptoms indicative of potential problems and a faculty for comprehending the combined effect of multiple forces. Naturally, most of the operational managers, used to the steady rhytm of normal operation and routine work lack the experience, training, competence and indication to control projects.
3. **Poor Control and Information System.** One of the factors, which effective control, is the poor quality of control and information system. Some of the weakness observed in control and information systems are:
   (*a*) **Delay in reporting Performance.** Often there is a delay in the reporting of performance. This prevents effective monitoring of the project and initiation of timely action to check adverse developments.
   (*b*) **Inappropriate Level of Detail.** Generally cost information for control is collected in terms of cost codes found in the company's cost accounting system. Irrespective of the level of detail employed for project planning and budgeting. Consider an extreme example wherein cost and value of work done are reported for the project as a whole. What is the value of such information for identifying where slippage is occurring and who is responsible for them.
4. **Unreliable Information.** One of the major problems in project control is unreliable and inaccurate data and information often project managers. Receive reports which suggest that "everything is okay" or things are "Reasonably within control" when the reality is otherwise. Further, for months after the project is completed, costs dribble into change a favourable variance into an unfavourable one or to aggravate an unfavourable variance.

Review Questions

1. What do you mean by coordination and control of a Project?
2. Explain Project Direction. What are the different stages of Project Direction?
3. Explain Project Communication with example.
4. What are the different problems in Communication? Explain the actions that can be taken to improve Communication.
5. What are the basic segments for a Project Management information system? Explain optimum Project Management information system.
6. What is Schedule Control? Explain different types of Schedule Control.
7. Explain Project Control and Cost Control.
8. What is the role of M.I.S. in Project Control.
9. Explain the reasons for in effective Project Control.

7 PERFORMANCE MEASURES IN PROJECT MANAGEMENT

Performance Indicators

A project will be considered totally successful. If it gets completed on time, within budget and perform exactly is the designer's specifications. A project, on the other hand, may be considered a total failure in the following cases:

1. If it is abandoned half way or kept in abeyance or completed with a changed concept.
2. If it does not produced as specified in terms of quality of produce.
3. If it becomes sick soon after going into commercial production.

In real life, a project cannot be considered either a total success or a total failure–it would fit somewhere in between. The invester, project manager and also the public who are watching the project are all keen to know how the project is being managed, and they must be given some indication about it while the project is still in progress:

Some performance indicators used are

1. **Time Over-run.** This basically means not sticking to the preplanned time schedule of the project. In order to get the project cleared through the approving authorities, a schedule may be fixed up which can be far from realistic. The vendors and contractors would only add to the confusion by promissing deliveries which can make anything possible on paper. Besides, a defective design and subsequent modification/change to suit the project requirements also increases the time and cost. This parameter, however, cannot be used as a true indicator for project management performance.

E.g.: A 200 tones per day mini cement project is quoted to have taken 12 to 24 months for completion. This time frame is not realistic as the gap is very large. Thus, as the exact time for completion is not available, the time over-run also cannot be accurately determined.

2. **Cost Over-run.** Anything done to a project including time over-run, would be reflected in the cost. If a project is not managed well its cost will go up and conversely if a project is managed well, its cost should come down. Therefore cost can be used as an indicator for project management performance. But, cost estimates in a project are to be revised at various stages to improve their accuracy, and they invariably increase after every revision. Cost over-run is used to represent the variance between the original sanctioned cost and final cost incurred.
3. **Project Sickness.** An efficient project manager is one who makes the best possible use of the resources given to him for achieving the project objectives. Whatever the project manger does with respect to the resources will be reflected in the cost and what the project manager provides in return to justify this cost is a plan which is to be called successful, must produce a saleable output. The ratio of this output to the cost incurred for putting up the plant could be an indicator of project management performance.
4. **Productivity as Performance Indicator.** A productivity indicator reflects how resources have been utilised either for production of goods and services or for creation of facilities for the same. Since a manager is responsible for utilisation of the resources put under his disposal, this indicator can reflect his performance. Productivity must be measured not merely for evaluating the performance of project management but mainly to ensure profitability of the plant and to ward of sickness. By working out the cost index and schedule index of a completed project, and comparing it with the industry average in the same technological area, it should be possible to get a true indication of the performance of the project management.
5. **Value as performance Indicator.** Value, in this context, indicates the quality of the products or services that are final outcome of the project. If the product or service does not meet the customers application needs then the ideas such as cost over-run, time over-run high productivity all become redundant.

Performance Improvement of CM and DM Company

The owner, when he is not doing things himself, needs someone. Who is not working for profit. There are the CM (Construction Management) and DM (Design Management) companies. These companies work as the owner's agents for a free and have only the owner's interest upper-most in their mind. They do not gain any profit from the sale of hardware. They earn a fee from the sale of their services and their fee represents a small percent of the project cost. Since they work for a fee, they have nothing to gain from an escalated project schedule or hardware cost. An owner's interest is, therefore, likely to be best served and the project cost is likely to be the lowest, when the project is handled by a competent CM or DM company.

CM Company

A CM company is specialised in the management of projects where field construction work constitutes the major part of the project work. Civil projects like housing, buildings, dams, highway etc. are the ones where CM companies have already established themselves.

Fig. 7.2 shows the project execution arrangement when a CM company is engaged. It shows that a CM company is not just for managing the construction phase of a project. It has to manage the entire project process covering planning, design, contracting and project implementation. However, the basic responsibility for planning and design rests with the architect-engineer, the CM company interacts with him for optimising project cost and avoiding construction problems. Likewise the basic responsibility for construction lies with the contractor. The CM company only undertakes the responsibility for construction, supervision co-ordination, and contract administration.

Project Management through CM Company—a Case

Hindustan Cement Limited (HCL) decided to use a CM company for constructing its township. It had also lined up a firm of architects (A/E firm) whose main responsibilities were to prepare the conceptual design, basic design package and all construction drawings. The A/E firm was also to approve payment schedules, change orders, provide job inspection and certify satisfactory completion.

The CM company was to provide construction expertise to A/E firm, assume full construction-phase responsibility and complete the township at a guaranteed maximum price. The CM company was to be paid 5% of this guaranteed maximum price as their fee. A/E firm was also to be paid an equivalent amount for its services.

Schedule, the CM team requested the A/E firm to develop the design phase schedule. This was later jointly finalised between the A/E firm and the CM company keeping the requirements of the master schedule in view in terms of priorities for release of building packages and construction drawings.

Construction Drawing Phase. During the phase the A/E firm was preparing the construction drawings and purchase specifications for equipments and work packages. It was however, the responsibility of the CM team to identify the work packages so that bid documents for procurement could be prepared accordingly. The work package plan was approved by the owner before forwarding the same to the A/E Firm. The CM company also identified the long delivery procurement items and indicated the priorities for release of bid packages in the A/E firm's design phase schedule.

The CM team also prepared a control estimate in terms of bid packages. This made it easier for the CM team to compare the prices quoted by the bidder with the control estimate. The various services provided by the CM team during this phase are given below:

1. **Preparation of approved contractor's list.** The CM team obtained the owner's approval for the contractors picked up from the list of prequalified contractors.
2. **Preparation of general conditions of control and obtaining owner's approval.** The CM team used their own standard contract form and obtained the owner's comments on the same.
3. **Preparation of tendering procedure.** The CM team modified their standard tendering procedure keeping the project location, schedule requirement and owner's organisation in mind is got the owner's approval of the same.
4. **Preparing tender document.** The CM team used the A/E's specifications and added general conditions of contract and bidding instructions.
5. **Bid evaluation.** The CM team evaluated the bids both technically and commercially. They carried out value engineering studies with the bidders to lower cost. In a number of cases the subcontractor's bid was converted to a cost plus contract with a fixed fee and the GMP.
6. **Negotiation and award of contract.** The CM team made recommendations to the owner for award of contract with or without further negotiation at the owner's end.

Construction Phase. During the construction phase the CM team took full responsibility for physical construction. They engaged the contractors and developed a detailed construction schedule for each contractor. The schedule was used as the basis for directing work, review of contractor's performance and expediting all work at site. The various services provided by the CM team during this phase are given below:

1. **Field Supervision.** The CM team deployed a well-staffed field organisation for day-to-day direction and co-ordination of the work of the various contractors, review of contractor's work progress, suggesting ways and means

for improving progress, identifying problems. To the owner after obtaining the A/E's approval.

2. **Value Engineering.** Search for better methods for construction continued even during the construction period, resulted in considerable savings in time and cost.
3. **Cash Requirement Forecasting.** Using detailed construction schedule and cost estimates, where design and procurement constitute the major part of the work, strength in design is considered essential for effective management. Given below is the presentation made by a design manage company to its prospective clients.

We are a design manage company. We design a project, we manage the project so designed, but what makes us different from others is that we manage through design.

Fig. 7.1, illustrates our mission, we design the hardware configuration which is referred to as scope, we design the framework within which the scope has to be contained and finally we detail our design and activities in such a way that a project configuration of scope-time-cost performance which the owner had approved is realised.

Design is our right hand, management is our left. We design with our own staff, and where we do not design we review and approve designs. But we do not have staff to manufacture fabricate or construct;

We only manage these activities. For this the owner enters into direct contract with vendors and contractors, as he enters into contract with us for design and management. However, he hands over the rains to use to manage the vendors and contractors. Fig. 7.1 show this arrangement.

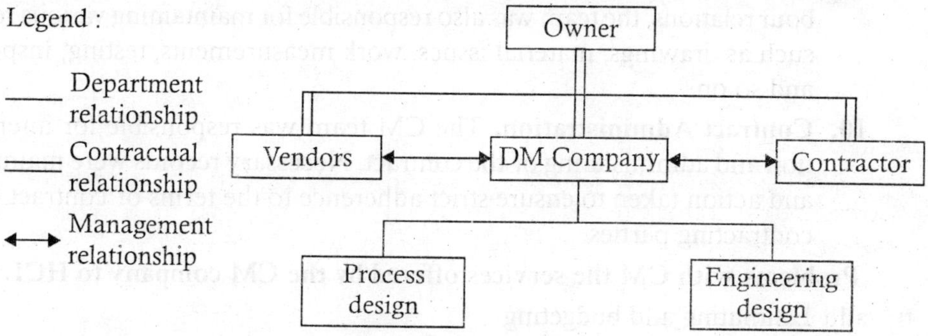

Fig. 7.1. DM Company.

Design is our forte. We consider it essential to manage and realise and high-tech project. Of the 2000 engineers we have, almost 1000 are engaged in process design and engineering. The complexity of technology and engineering requires not-only a high technical specialisation but also a tremendous amount of co-ordination on a day-to-day basis. It is, therefore, essential to carry this specialisation in-house so as to be able to deliver a top class plant in minimum time and a least cost.

Organisation: our entire company rallies behind a project manager for the execution of any project. The project manager act as our interface with the outside world. He interacts with the owner, the vendors, contractor, government agencies whoever requires to be interacted forecast were provided to the owner regarding requirement of each to pay running account bills.

4. **Inspection of Work.** The CM team jointly with the A/E carried out inspection to ensure adherence to drawings and specifications provided by the A/E all completed work was, however, finally approved by the A/E.
5. **Shop Drawings.** The CM team expedited the release of shop drawings from vendors and contractors and their approval through the A/E.
6. **Change Orders.** The CM team processed all change requests involving the A/E where needed and made recommendations to the owner for acceptance.
7. **Extra Claims.** The CM team processed all extra claims and made recommendations to the owners.
8. **Warehouse Management.** The CM team took complete responsibility for storage, issue, record and reconciliation of owner-supply materials to various contractors.
9. **Safety, Job Site Records and Labour Relations.** The CM team was responsible for maintaining the safety programme and also harmonious labour relations, the team was also responsible for maintaining various records such as drawings, material issues, work measurements, testing, inspection and so on.
10. **Contract Administration.** The CM team was responsible for interpretation and administering of the contract. Necessary records were maintained and action taken to ensure strict adherence to the terms of contract by the contracting parties.

Problems with CM the services offered by the CM company to HCL were:

1. Estimating and budgeting.
2. Value engineering.
3. Bid packaging.
4. Bid analysis and contract negotiation.
5. Supervision and co-ordination.

6. Scheduling and cost monitoring.
7. Performance and quality control.

The above services, as one can see, are all geared to ensure completion of the project at the least cost without restricting the freedom of the A/E firm to create a design that best meets the owner's requirement. Thus, with the CM company owners gets all the advantages of scientific project management.

Unfortunately, things may not work as they should particularly if there is no good rapport between the A/E firm and the CM team. The former may not respond to the suggestions of the CM team for design modifications which could reduce project cost as it may feel that thin will lower its in the eyes of the owner. The vendor may also not come forward with a cost-effective design under the CM team proposes one. This means that specialised knowledge in construction may not be enough for delivering a project at least cost.

DM Company

Due to the above mentioned shortcomings, owners look for those companies who have competence of design besides expertise in construction for the execution of their projects. Such companies are known as design manage companies. For managing complex hi-tech projects.

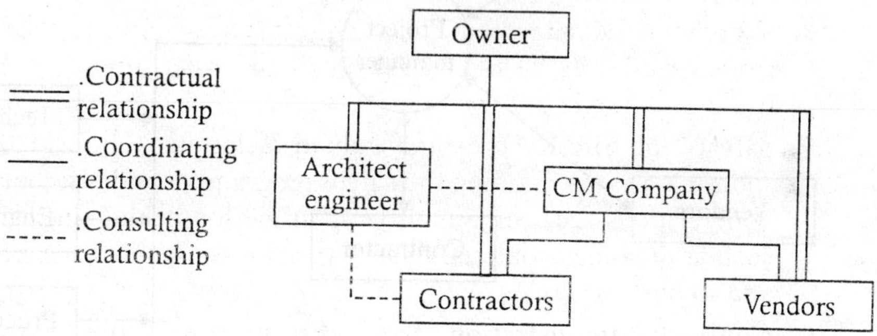

Fig. 7.2. CM Company.

Conceptual Design Phase. During this phase the A/E firm developed block diagrams that would determine the size and shape, square footage and various functional spaces. As the A/E firm completed the functional programme for each part of the project, the CM company prepared an estimate of cost and compared it with the budget estimate. The estimate provided by the CM company their requirements for what was considered in the feasibility report. Necessary adjustments were made to bring the project back within the budget.

Basic Engineering Phase. During this phase the A/E firm prepared preliminary floor plans for each type of quarter. A/E firm's consultant civil, mechanical and electrical engineers, prepared basic designs documents for utility systems and various services. The CM company prepared system estimates to identify where the money is being spent and attached the problem areas through value engineering analyses. The CM team brought in sub-contractors as consultants to supplement its in-house capabilities. The team made a number of recommendations resulted in considerable saving for the owner.

As the A/E firm developed building layout and space relationship. The CM team studied the time and cost savings. At the end of these studies the GMP was finalised. The CM team also developed a master schedule at this stage in close co-ordination with the owner and A/E firm, keeping the financial position in full view. Based on this master schedule for the execution of a project. He is provided all the necessary support by the company. Fig. 7.3 shows this arrangement.

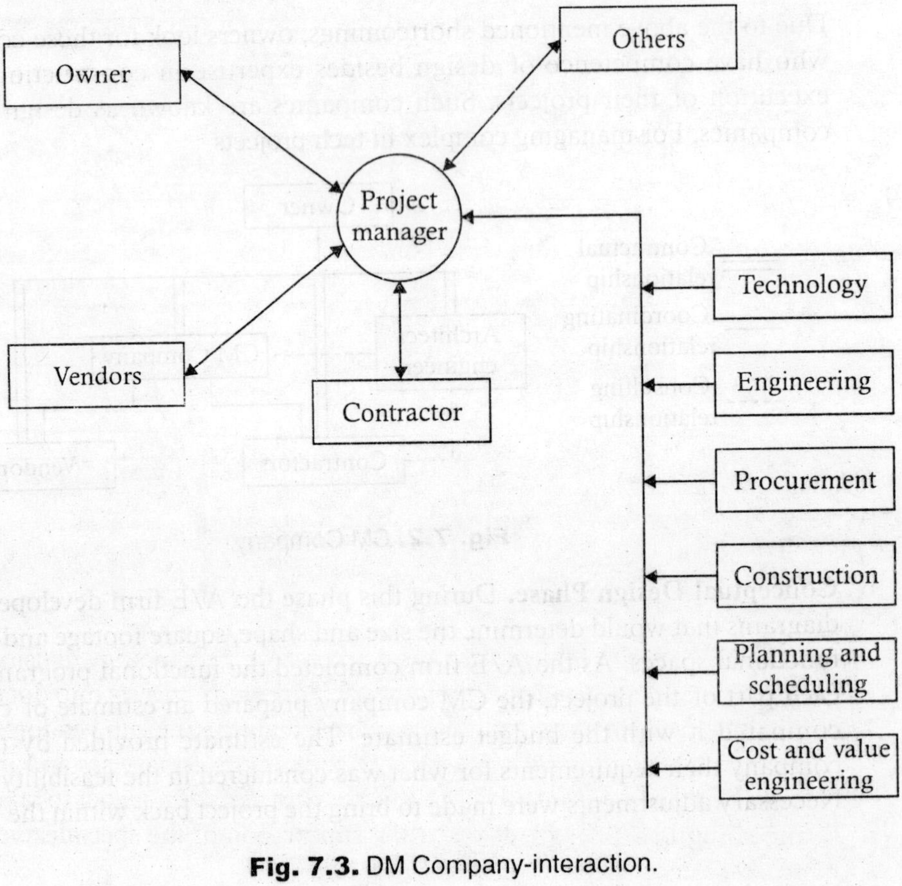

Fig. 7.3. DM Company-interaction.

"We operate on a matrix. The project managers make requisitions. For the help he needs and the functional departments provide him the services. The project manger decides what is to be done, when it is to be done and at what cost. The functional manager decides how the work is to be done to meet the project manager's requirements as also the functional requirement. An individual functional staff thus, has two bosses–a project boss and his own functional boss. Inspite of this we face no problems. We have been operating in this way almost from the inception and have got used to it.

Figure 7.4 Shows the Arrangement

"We take total responsibility for the execution of a project, from concept to commissioning, as the owner's agent for a small fee which could be fixed as a percent of the total project cost or a lumpsum figure related only to our effort and not to the total project cost. We can also be reimbursed for our effort at actuals plus a fixed fee. Thus, while we take a turnkey responsibility we do not work for profit as a turn key contractor would do. We can also guarantee a maximum price. Which is normally the reason why some owners opt for a turn-key contract, at an appropriate stage of the project.

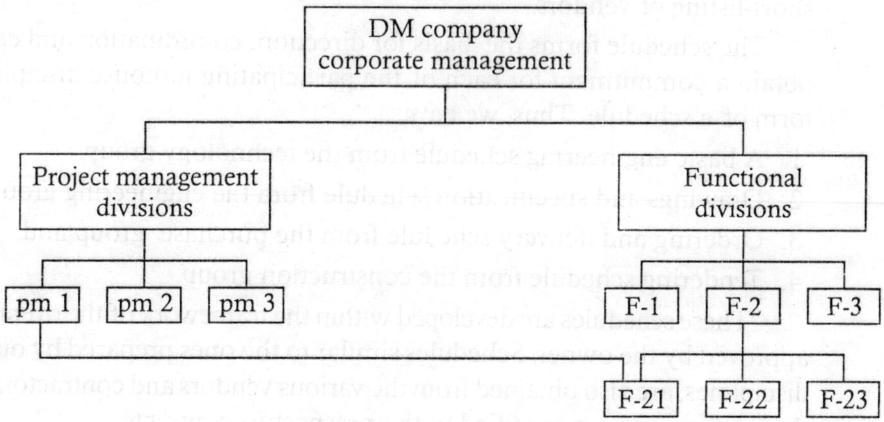

Fig. 7.4. DM Company-matrix.

Method of Project Execution

Fig. 7.5. Explains how we execute a project. Our technology division provides the basic packages which the engineering division converts into documents for purchase and construction. The purchase division lines up vendors. Similarly, the construction division lines up the contractors, supervises and inspects the work being carried out by the contractors, we assist the owner to take trial runs and commission the plant under the supervision of the specialist vendors and process licensors, as the case may require.

"In all projects the owner's requirement in our starting point the owner first defines the scope of work and provides the design basis. Using the owner's design basis and our know-how or the process licensor's know-how our process division prepares the basic package for the project. The basic package when approved by the owner becomes the bible for the project.

We consider this as the zero date for the project as this enables us to fix the project completion date realistically.

"But before the zero date out, cost-cutting effort will be almost at its peak. We evaluate the owner's design basis for its cost and worth and work our alternative processing schemes that can ensure better value for the project. Since our value engineering team make us look at several alternatives, we may take a little time in finalising the basic package. But since the opportunity for cost control is maximum during this stage, we do not want to trade this opportunity for a little saving on time which we will make up by avoiding changes at a later date. We seek our client's indulgence for this.

"Almost simultaneously we require our client's participation for finalising the work packages, list of contractors, purchase division list, purchase procedure, short-listing of vendors.

The schedule forms the basis for direction, co-ordination and control. We obtain a commitment for each of the participating in-house disciplines in the form of a schedule. Thus, we have:

1. A basic engineering schedule from the technology group.
2. Drawings and specification schedule from the engineering group.
3. Ordering and delivery schedule from the purchase group and
4. Tendering schedule from the construction group.

"These schedules are developed within the framework of the overall schedule approved by the owner. Schedules similar to the ones prepared by our in-house disciplines, are also obtained from the various vendors and contractors for which the requirement are specified in their respective contracts.

"We emphasize co-ordination, communication and team work. Our project management team for a major project includes project co-ordinators, an engineering co-ordinator, a procurement co-ordinator ensures the co-ordination of all aspects of work for the project units under him; the engineering co-ordinator, co-ordinate the work of various engineering disciplines, the procurement co-ordinator co-ordinates the work of purchasers, vendors, inspectors and traffic staff; the construction co-ordinator co-ordinates the release of drawings and materials to meet the requirements of construction and the scheduling co-ordinator co-ordinates the work of all these co-ordinators and works for the project manager. We also hold regular co-ordination meetings with our in house staff, vendors and contractors.

"We control progress through continuous monitoring, expediting and follow-up. We hold project-level review meetings at home at the office and at field. In these meetings the actual progress is compared with the schedule and the adequacy of inputs and work front for giving scheduled output is reviewed. Commitments are taken from various groups regarding inputs, workfront and outputs for the next reviewing period. The commitment are followed up over the telephone, through letter of personally till they are actually fulfilled.

"We send a regular monthly report to our client indicating accomplishments, completion, time, budget forecasts and areas of concern. The owner often calls us for a meeting to discuss this report and work out new strategies where required. Thus, we take our client with us all the time and keep him informed of all developments in the project.

"These reports also reach our corporate management executives who are, thus, kept informed about the developments of a project. But the matter does not end here. We have a project management audit group which presents the project performance report used by the corporate management for reviewing the project at every quarter. The project management audit group analyses the project performance independent of the project manager and thus enable both the projects as well as the corporate management to get an unbiased view of the project status. The audit report also enables a project manager to see his team's efficiency which he would not have known through routine status reports. The corporate review also helps the project manager to involve the corporate management on matters where high level intervention may become necessary for successful completion of a project.

"Thus when you entrust your project to us, you will find the entire organisation working for you with the sole intention of seeing the project completed in the way you would have liked to do it yourself".

And contractors, co-ordination procedure, overall schedule, budget and overall project layout.

This we call our project execution strategy, and we insist on our client's fullest participation in finalisation of the same. Once the basic package and the project execution strategies are thus, jointly finalised, we proceed with full steam.

"For project startup, we hold what we call (project kick–of meeting). We call all our functional managers to nominate the lead engineers. The meeting discusses the project scope, procedures and execution strategies with the functional managers and the lead discipline engineers."

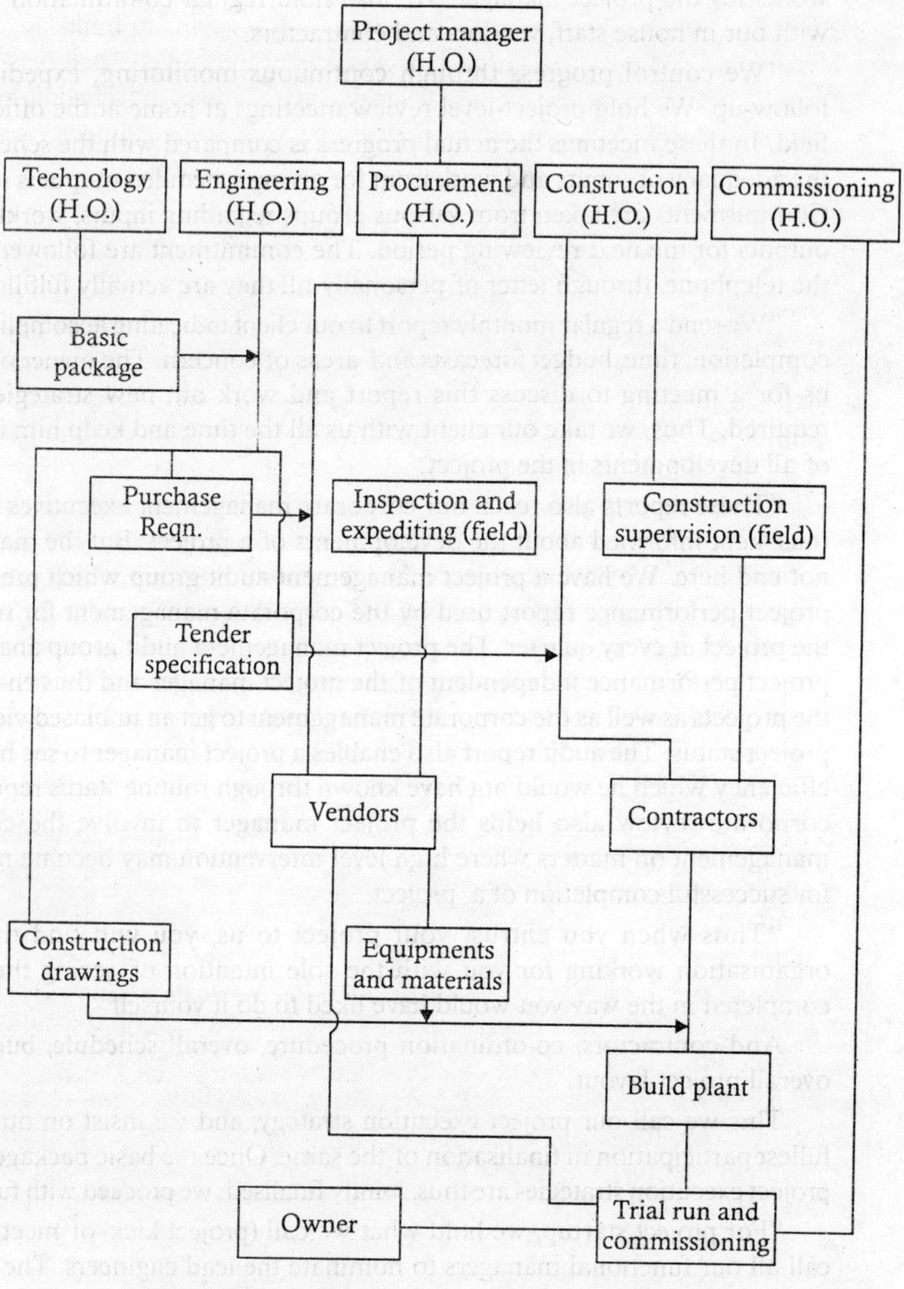

Fig. 7.5. DM Company-project execution system.

The project manager also indicates how he plans to build up the project team and seeks the support of the functional managers. Each lead discipline engineer is asked to prepare his job list and then develop the schedule he proposes to follow with the assistance of the project scheduling engineer.

The Problems with DM Company

The real world, unfortunately, is not as beautiful as any presentation would try to make it. The best intentions are often at the mercy of situations. A DM company may not show profit if the cost of the project hardware goes up, but it has to certainly put a lot of man hours on value engineering to reduce cost of the hardware. The question, therefore, is, will the company put this additional effort for no financial gain? If the DM company is paid a lumpsum fee for design and management, will the company not try to reduce effort in design? Will the company not try to use previous designs and reduce managerial effort in design? Will the company promise to work for the owner? Will the company not direct its resources and slacken an on-going job in order to take up another one? Will the company not try to pass on its delays in engineering and procurements to vendors and contractors when it comes to payment of penalties? Thus, what may sound ideal at the hypothetical level, may not necessarily prove so in reality because of many compulsions in the real word.

Project Management Environment

Project management performance will largely depend on the real-world environment. The project management environment in India, is very different from any other country. There are many problems which are peculiar to our country and these are experienced by all those who are concerned in the execution of both small and big projects. One has to be aware of these problems in order to be able to cope with the same for successful implementation of a project.

The most important problem is lack of mutual trust and respect amongst the participating agencies: owner, financial institutions, consultants, vendors and contractors. The owner believes that the agencies/contractors would take him for a ride and, therefore, he should, as far as possible, do things himself. When consultants are not appointed, projects are likely to have congenital weaknesses such as wrong selection of technology, wrong site, high risk element etc. Sometimes the owner may appoint a consultant for a normal fee and ask him to prepare a report which he can sell to the bank. These reports often do not reflect reality as they are made without any in -depth study, and if cleared, would give birth to defective projects. This, no doubt, reflects on a consultants lack of professional ethics and can be avoided if the financial institutions use a proper accreditation, system for consultants.

However, accreditation of consultants may not set everything right. A site may often be selected purely on personal rather than on techno-economic considerations. The same may happen with the selection of technology of even with the selection of the consultant. Such projects would certainly run into rough weather irrespective of whether a consultant or CM/DM company is engaged or not.

A competent consultant or a CM/DM company can also not help when the promoter wants to have short-term gains at the cost of the project. This may happen with traders turned entrepreneurs or with those who are basically financiers. The pressure that such an entrepreneurs or with those who are basically financiers. The pressures that such an entrepreneurs may bring on a consultant, vendor and contractor would invariably make the project environment sick. To avoid this it is often suggested that besides technical and financial appraisal of a project the financial institutions should appraise the entrepreneur himself. It is also suggested that the financial institutions should introduce an on-going audit system to prevent diversion of funds and other forms of financial irregularities.

Review Questions

1. What are the performance measures to be adopted in Project Management? Explain them.
2. Explain performance indicators in detail.
3. Explain the steps involved in performance improvement of CM and DM Company.
4. Explain the Project Management through CM Company—A case.
5. With a neat sketch explain the DM Company Interactions.
6. Explain the DM Company Matrix.
7. Explain the method of Project Execution.
8. Explain the Project Management Environment.

8 CASE STUDIES ON PROJECT MANAGEMENT

PROJECT OVER-RUN

The green company production project was completed three months behind schedule and at a cost over-run of approximately 60 percent. Following submittal of the final report, Phil Graham, the director of project management, called a meeting to discuss the problems encountered on the green project.

Phil Graham : "We're not here to point the finger at anyone. We're here to analyse what went wrong and to see if we can develop any policies and/or procedures that will prevent this from happening in the future. What went wrong ?"

Project Manager : "When we accepted the contract, Green did not have a fixed delivery schedule for us to go by because they weren't sure when their new production plant would be ready to begin production activities. So, we estimated 3,000 units per month for months five through twelve of the project. When the found that the production plant would be available two months ahead of schedule, they asked us to accelerate our production activities. So we put all of our production people on overtime in order to satisfy their schedule. This was our mistake, because we accepted as fixed delivery data and budget before we understood everything".

Functional Manager: "Our problem was that the customer could not provide us with a fixed set of specifications, because the final set of specifications depend upon the OSHA and EPA requirements which could not be confirmed until initial testing of the new plants. Our people, therefore, were asked to commit to man-hours before specifications could be reviewed.

"Six months after project go-ahead, Green company issued the final specifications. We had to remake 6,000 production units because they did not live up to the specifications".

Project Manager : "The customer was willing to pay for the remake units. This was established in the contract. Unfortunately, our contract people didn't tell me that we were still liable for the penalty payments if we didn't adhere to the original schedule".

Phil Graham : "Don't you feel that misinterpretation of the terms and conditions is your responsibility?"

Project Manager : "I guess I'll have to take some of the blame".

Functional Manager: "We need specific documentation on what to do in case of specification changes. I don't think that our people realize that user approval of specifications is not a contract agreed to in blood. Specifications can change, even in the middle of a project. Our people must understand that, as well as the necessary procedures for implementing change".

Phil Graham : "I've heard that the functional employees on the assembly line are grumbling about the green project. What's their gripe?"

Functional Manager: "We were directed to cut out all overtime on all projects. But when the green project got into trouble, overtime became a way of life. For nine months, the functional employees on the green project had as much overtime as they wanted. This made the functional employees on other projects very unhappy.

"To make matters worse, the functional employees got used to a big take-home pay check and started living beyond their means. When the project ended, so did their overtime. Now, they claim that we should

		give them the opportunity for more over-time every body hate us".
Phil Graham	:	"Well, now we know the causes of the problem. Any recommendations for cures and future prevention activities?"
Gray	:	"Yes, I received your memos. But as I told you before, I have enough problems to worry about without doing four job for you. If you can't handle the work let me know and I'll find some one who can".

Eddie returned to his desk and contemplated his situation. Finally, he made a decision. Next week he was going to put a signature block under his for Gary to sign, with carbon copies for all division managers. "Now, let's see what happens", remarked Eddie.

Case Study: The Two-Boss Problem

On March 1965 Martin was assigned full time to project turnbolt by Fared manager of the thermodynamics department. All work went smoothly for four and one-half of the five months necessary to complete this effort. During this period of successful performance Martin had good working relation with Esmail (The turnbolt project engineer) and Fared.

Fared treated Martin as a theory. Y. Employee. Once a week Fared and Martin would chat about the status of Martin work. Fared would always conclude their brief meeting with, "you're doing a fine job, Martin. Keep it up. Do anything you have to do to finish the project".

During the last month of the project Martin, began receiving conflicting requests from the project office and the department manager as to the preparation of the final report. Compton told Martin that the final report was to be assembled in view graph format (*i.e.*, "bullet" charts) for presentation to the customer at the next technical interchange meeting. The project did not have the funding necessary for a comprehensive engineering report.

The thermodynamics department, on the other hand, had a policy that all engineering work done on new projects would be documented in a full and comprehensive report. This new policy was implemented about one year ago when Fared become department manager. Rumor had it that Fared wanted formal reports so that he could put his name on them and either publish or present them at technical meetings. All work performed in the thermodynamics department required.

Fared's signature before it could be released to the project office as an official company position. Maintain a large editional or graphic arts department. Personnel designing to publish had to get the department manager's approval and, upon approval, had to prepare the entire report themselves. Without any "over head" help. Since Taylor had taken over the reins as department head he had presented three papers at technical meetings.

A meeting was held between Martin, Fared and Esmail.

Esmail : "I don't understand why we have a problem? all the project office wants is a simple summary of the results. Why should we have to pay for a report that we don't want or need?"

Fared : "We have professional standards in this department. All work that goes out must be fully documented for future use. I purposely require that my signature be attached to all communications learning this department. This way we obtain uniformity and standardization. You project people must understand that, although you can institute your own project policies and procedures (within the constraints and limitations of company policies and procedures). We department personnel also have standards. Your work must be prepared within our standards and specifications".

Esmail : "The project office controls the purse strings. We (the project office) specified that only a survey report was necessary. Further more if you want a more comprehensive report, then you had best do it on your own over head account. The project office isn't going to foot the bill for your publications."

Fared : "The customary procedures is to specify in the program plan the type of report requested from the departments. In as much as your program plan does not specify this. I used my own discretion as to what I thought you meant".

Esmail : "But I told Martin what type of report I wanted. Didn't he tell you?"

Fared : "I guess I interpreted the request a little differently from what you had intended. Perhaps we should establish a new policy that all program plans must specify reporting requirements. This would alleviate some of the misunderstandings, especially since my departments has several projects going on at one time in addition I am going to establish a policy for my department that all requests for interior, status on final reports be given to me directly. I'll take personal charge of all reports".

Esmail : "That's fine with me ! and for your first request. I'm giving you an order that I want a survey report, not a detailed effort."

Martin : "Well since the meeting is over, I guess I'll return to my office (and begin updating my resume just in case)".

Case Study: Scheduling the Safety Lab

Dr. Nair : Director of Research laboratory remarked Ram "Now see here, Ram I understand your problem well" "I pay you a good salary to run the safety labs. That salary also includes doing the necessary scheduling to match our priority. Now, if you can't handle the job, I'll get some one who can".

Ram : "Every Friday morning your secretary hands me a sheet with the listing of priorities for the following week. Once just once, I'd like to sit in on the director's meeting and tell you people what you do to us in the safety lab when you continually shuffle around the priorities from week to week."

"On Friday afternoon, my people and I meet with representatives from each project to establish the following week's schedules".

Dr. Nair : "Can't you people come to an agreement?"

Ram : "I don't think you appreciate my problem. Two months ago, we all sat down to work out the lab schedule. Project X-13 had signed up to use the lab last week. Now, mind you, they had been scheduled for the past two months. But the Friday before they were to use it, four new priority list forced them to reschedule the lab at a later date, so that we could give the use of the lab to a higher-priority project. We're paying an awful lot of money for idle time and the redoing of network schedules. Only the project managers on the top-priority projects end up smiling after our Friday meetings".

Dr. Nair : "As I see your problem, you can't match long-range planning with the current priority list. I agree that it does create conflicts for you. But you have to remember that we, have many other conflicts to resolve I want that one solved at your level, not mine".

Ram : "Every project we have requires use of the safety lab. This is the basis for our problem. Would you consider lettering us modify your priority list with regard to the safety lab?"

Dr. Nair : "Yes but you had better have the agreement of all of the project managers. I don't want them coming to see me about your scheduling problems".

Ram : "How about if I let people do long-range scheduling for the lab, say for three out of the four weeks each month? The fourth week will be for the priority projects".

Dr. Nair : "That might work, you had better make sure that each project manger informs you immediately, of any schedule slippages so that you can reschedule accordingly. From what I've heard, some of the project managers don't let you know until the last minute".

Ram : "That has been part of the problem. Just to give you an example, project VX–161 was a top-priority effort and had the lab schedule by two weeks. They walked into my office and demanded use of the lab for the third week in February. Since they had the top priority, I had to grant them their request. However, project BP-3 was planning on using the lab during that week and was bumped back three weeks. That cost then a pile of bucks in idle time pay and of course, they're planning me".

Dr. Nair : "Well Ram, I'm sure you'll find a solution to your problem."

Case Study : Tristar International

On June 15 1964, the department of energy resources awarded Tristar a Rs. 475,000 contract for the developing and testing of two waste treatment plants. Tristar had spent the better part of the last two years developing waste treatment technology under their own R and D activities. This new contract would give Tristar the opportunity to "break into a new field" that of waste treatment.

The contract was negotiated at a firm fixed price. Any cost over-runs would have to be incurred by Tristar. The original bid was priced out at Rs. 847,000. Telestar's managemnt, however, wanted to win this one. The decision was made that telestar would "buy in" at Rs. 475,000 so that they could at least get their foot into the new market place.

The original estimate of Rs. 847,000 was very "rough" because Tristar did not have any good man-hour standards in the area of waste treatment, upon which to base their man-hour projections. Corporate management was willing to spend up to Rs. 400,000 of their own funds in order to compensate of bid of Rs. 475,000.

By March 19 ,1965 cost were increasing to such a point where over-run would be occurring well ahead of schedule. Anticipated costs to completion were now Rs. 943,000. The project manager decided to stop all activities in certain functional departments, one of which was structural analysis. The manager of the structural analysis department strongly opposed the closing out of the work order prior to the testing of the first plant's high-pressure pneumatic and electrical systems.

Structures Manager : You're running a risk if you close out this work order. How will you know if the hardware can withstand the stresses that will be imposed during the test? After all, the test is scheduled for next month and I can probably finish the analysis by then".

Project Manager : "I understand your concern but I cannot risk a cost over-run. My boss expects me to do the work within cost. The plant design is similar to one that we have tested before, without any structural problems being detected. On this basis I consider your analysis as unnecessary".

Structures Manager : Just because two plants are similar does not mean that they will be identical in performance there can be major structural deficiencies".

Project Manager : "I guess the risk is mine.".

Structures Manager : "Yes, but I get concerned when a failure can reflect upon the integrity of my department. You know, we are performing on schedule and within the time and money budgeted. You're setting a bad example by cutting off our budget without any real justification".

Project Manager : "I understand your concern, but we must put out all stops when over-run costs are inevitable".

Structures Manager : There's no question in my mind that this analysis should be completed. However, I'm not going to complete it on my overhead budget. I'll reassign my people tomorrow. Incidentally, you had better be careful. My people are not very happy to work for a project that can be cancelled immediately. I may have trouble getting volunteers next time".

Project Manager : "Well, I'm sure you'll be able to adequately handle any future work. I'll report to may boss that, I have issued a work stoppage order to your department".

During the next month's test, the plant exploded. Post analysis indicated that the failure was due to a structural deficiency.

(*a*) Who is at fault?

(*b*) Should the structures manager have been dedicated enough to continue the work on his own?

(*c*) Can a functional manager, who considers his organization as strictly support, still be dedicated to total project success?

Case Study: The Problem with Priorities

For the past several years Sudha Corporation had achieved remarkable success in winning R and D contracts the customers were pleased with the analytical capabilities of the R and D staff at Sudha Corporation. Theoretical and experimental results here usually within 95 percent agreement. But many customers still felt that 95 percent was too low. They wanted 98.99% (percent).

In 1973 Sudha updated their computer facility by renting on IBM–370 computer the increased performance with the new computer encouraged the R and D group to attempt to convert from two-dimensional to three dimensional solutions to their the oretical problems. Almost everyone except the director of R and D thought that this would give better comparison between experimental and theoretical data. Sudha Corporation had tried to develop the computer program for three-dimensional solutions with their own internal R and D programs, but the cost was too great. Finally after a year of writing proposals, Sudha Corporation convinced the federal government to sponsor the project. The project was estimated at Rs. 7,50,000 to begin 9th April 1995 and to be completed by 19th March 1996 Rajesh was selected as project manager Raj had worked with the EDP department on other projects and knew the people and the man-hour standards.

Sudha Corporation was big enough to support one hundred simultaneous projects with so many projects in existence at one time, continual reshuffling of resources was necessary. The corporation directors met every Monday morning to establish project priorities. Priorities were not enforced unless project and functional managers could not agree upon the allocation and distribution of resources.

Because of the R and D director's persistence the computer project was given a low priority. This posed a problem for Raj. The computer department manager refused to staff the project with his best people. As a result Raj had severe scepticism about the success of the project.

In July two other projects manager held a meeting with Raj to discuss the availability of the new computer model.

"We have two proposals which we're favoured to win, providing the we can state in our proposal that we have this new computer model available for use" remarked one of the project managers.

"We have a low priority and, even if we finish the job on time, I'm not sure of the quality of work because of the people we have assigned" said Raj

"How do you propose we improve our position?" asked a project manager.

"Let's try to get in to see the director of R and D" asserted Raj.

"And what are we going to say in our defence?" asked one of the project managers.

Three project managers with distinctly different roles

1. **The site manager of a housing development.** 'I am in charge of the construction of the buildings you see around you (he gestured with his hand to the mixture of partially and fully completed properties) and of making sure they go from this stage (he indicates a pile of drawings and building schedules) to the point where we can hand them after to the sales people to sell. Most of the work is supervisory, ensuring that orders are placed and materials arrive on time, people turnup, do the job property and get paid for it at the end of the weak. There are always arguments between the various trades–people to resolve and problems just get dumped on the desk. Some of the toughest problems come with the people you have to work with. Some of them will do anything to try to get one over on you–they'll tell you a job is finished when you can see it is only half done. Unless you go and check it yourself you're in trouble also, they don't give a damn for my schedule. How do you get a roofer, at four O'clock in the afternoon when it is raining rather heavily (not the words actually used) when you know he has a long drive home, to get back on the roof and finish the job he is doing so that other jobs which self on this being completed can start at eight O'clock the following morning? It wouldn't be the first time we had to block his car in with a parallel of bricks to stop him leaving'.
2. **Implementing total quality management the quality director.** "The quality director was appointed with the brief to introduce total quality management (TQM) to the company. It was his responsibilities to put the proposal as to how it could be done, and then to carry it out. As he described at the outset of the project" (this) is one of the most complex projects that we could undertake at this time". The complexity came because the project would hopefully change the way that everyone in the company thought and worked (*i.e.*, both attitude and procedures). This would have to be done through consultation, training and the demonstration through plotting small-scale improvement activities, that the move towards TQ was worthwhile. The initial phase as part of the proposal process was to carry out a company-wide quality audit to determine attitudes, knowledge and current practice. The results paved the way for the carrying out of targeted efforts where

needed most. The first phase of execution was to take the board of directors of the company on-awareness training-showing them how working under a TQ environment would benefit them, and what changes would be needed. The next level of management level. These managers then trained their own people–a process known as "cascading". The project to introduce the new philosophy to the company took several years, and has now moved on to become an accepted way of working. The quality director was initially involved in the management of the introduction process, where the employees and suppliers needed to be convinced that this was a good route for the company to take. His role then became one of project sponcer of a variety of improvement projects, which may be considered as sub-projects of the main-one'.

3. **Project manager is financial management system implementation.** The main roles of the job include:

- Organisation-from the design of the system to determining support issues and providing training;
- Anticipation of future requirements of the system;
- Monitoring of progress of the implementation;
- Communication and information providing progress reports to local team members and national common interest groups;
- Audit-ensuring the house keeping, procedures and system security are in order.

The initial system design work involved co-ordinating with external system designers, the providers of the software and the in-house IT group. Our local area network (LAN) needed up grading to run the new system. Other organisational issues were the role that consultants would play in the system design and training of users and the allocation of the budget between activities.

Anticipation was required as the requirements of the system would change over its life. For example higher level monthly indicators of financial performance would need to be provided where they had not been needed before. If addition, a management accounting system would be needed to provide budgetary controls.

The monitoring system we used for the project was PRINCE. This provided a basic set of planning tools and we filled in the blanks on the planning sheets.

A team was setup to monitor progress against the plan.

Training was one area where I was personally involved with the users, showing them how to use the system. People are very frightened of technology and do not always group immediately ideas you think are very simple. This is where the greatest attribute of the project manager was needed in plenty-patience'.

Case Study: The Channel Tunnel Project

Several years late and coming in at double the initial budget (around £ 10 billion as opposed to the estimated £ 5 billion) set for the project, the channel tunnel was opened to passenger and freight traffic during 1994. At several times during the work, the company which was funding the tunnelling work (Eurotunnel) itself ran out of money and had to return to the stock market and its banking financiers to seek new funds. The openings of the various services, attended by the world's press, were not without hitch either. Trains 'failed to proceed' at the times indicated, and had to be replaced, and there were well-publicised lapses in security arrangements by the operation staff. Tales were recounted of how the tunnel, dug simultaneously from the French and English sides, was intended to meet in the middle, but did not. Add to this the loss of life during the project (ten people were killed during the tunnel construction), and the ongoing debate as to the desirability of the high-speed rail-link through Kent (trains are environmentally friendly until they are routed through your back garden), and the general public would be forgiven for thinking that the project was very badly managed. They have short memories, however, and the media focus has shifted to the way in which the business of transporting people, vehicles and freight is run.

The station at waterloo, the terminal in Kent and the French stations have had to be constructed too, but have gained for less publicity. The waterloo Eurostar station has won various awards for its architecture as well as praise from users who are more used to the draughty, dingy stations of the UK-rail network.

By the start of 1995, the service being provided was getting good press reports, and the passenger transport side of the business was running ahead of predictions. Many of the earlier operating problems appear to have been rectified through when one train was cancelled and the passengers offered alternative transport on the ferries, the tannoy announcement by the captain of the ship taking them did have more than a little of sense of gloating about it ! The air lines and ferry companies have not been slow to respond–the entrance of another player into a highly competitive market has been unwelcome to the industry, through not unbeneficial to the customers.

The Trains

The Eurostar service is the passenger carrier, with each of the high-tech trains carrying more than the equivalent of two jumbo-jet loads of passengers at one time (over 800). It is owned by a consortium of British, Belgian and French railways. The trains themselves, over a quarter of a mile long, are fitted out so that they resemble the interior of an aircraft rather then a train. The chairs are reclining and individual reading lights are provided.

The train can travel at 186 mph (289 kph) on specially adapted tracks in northern France and around 100 mph.

Through Kent. The engines for the trains were specially developed by GEC–Alsthom, an Anglo-French company, and can run on the four different power systems. That the train will come across–French, Belgian, British and the tunnel itself. Much of the rest of the train design comes from the highly successful French traines–the TGV (Train a grande vitesse).

Le shuttle is the part of the service for cars, caravans and coaches–the box like carriages transport the vehicles with their occupants remaining within the vicinity of their vehicles during the journey, currently from foucestone to calais. The completion for the ferries is that this service is virtually weather proof and only takes 35 minutes from platform to platform.

The safety systems against one of the greatest fears of all-five-are where the greatest application of technology has occurred. The five breaks, detection systems and procedures for dealing with an outbreak are way in excess of anything previously related to a train. The freight service for taking lorries and containers through the tunnel runs on different trains to the above services.

The Tunnel

At the outset of the project, the technology involved and the sheer scale of the tunnel digging meant that it would not be possible for it to be undertaken by one company. A consortium of ten companies (5 French and 5 British) was formed and had to co-ordinate the digging of 3 tunnels (one each way and one service tunnel) as well as the 1200 suppliers of 3 tunnels (one each way and one service tunnel) as well as the 1200 suppliers and 15000 workers. Once this had been achieved, the tracks (over 130 miles in total) were laid, and the control system for the running of the trains installed. From an engineeing point of view, the sheer scale of the project and the fact that it has been achieved at all shows the project to be an immense success.

The project appears to have achieved what is set out to achieve–there is now a fixed transport link between the UK and the rest of Europe, one that it is hoped will encourage the kind of union that many feel is needed. If Europe is to continue its economic growth.

Case Discussion

1. Identify the major components of the channel Tunnel project.
2. Distinguish between the project management and the line management functions which have been described here.
3. Describe the complexity of the project.
4. Discuss the issues that would make you believe that the project has been a success or failure.

Case Study: Implementing BS-EN-ISO 9000 in a Health Service Environment

The international standard for quality systems BS-EN-ISO 9000 define the way in which an organisation should set its own standards and provide procedures for reaching those standards. It concerns the way that an organisation operates rather than the standards of the output itself (products, services, etc). The environment in which the unit was operating was that of intense financial pressure, coupled with the indication that the functions of the unit would be put out to commercial tender in the near future (through market-testing). It was also hoped that the discussion of quality and the formalisation of parts of the organisation would give accountability and help focus on customer needs.

The move to bring their method of operating within the requirements of the standard came as part of the measures to give them an advantage in any future competitive tendering situation for health services. The project was set a target of achieving the necessary certification with two years. In some areas and longer in others. Staff would need to be trained and the necessary documentation (detailing all working procedures) would have to be prepared.

The project was started and a new project manager (called the quality manager) recruited to provide both knowledge and a degree of independence from the other functions within the organisation. Across the organisation, there were pockets of enthusiasm, but otherwise annoyance from departmental managers who saw the changes as being bureaucratic and as interference from outsiders in the way they ran their departments. Above all, there was resistance to change which was both obstructive and destructive. The common feeling was that people had enough to do already, and this was an additional workload.

The result of this was that the areas, which took the new standard on board, and made it work in their way, got through the process earliest and with greatest benefit. On reviewing the implementation, it was shown that one of the reasons that people had opposed the changes was a lack of comprehension as to what they would mean in practice. They could not see how the documentation of adherence to and continuous improvement of working processes would improve their lives, or improve the organisation.

Case Discussion

1. What was the need and how did the project go towards meeting this?
2. Describe the complexity of the project.
3. How might identifying all the people who could hold up progress earlier in the project have increased the speed of achieving the project goals?
4. Why might recruiting an external project manager be beneficial, and what risks would there be?

The requirements of a successful project managers (PM)

The role that the PM will play is determined by:

- The nature of the project–complexity, scale, position in hierarchy of projects,
- The nature of the organisation in which it is being carried out, *e.g.*, sector, activities, organisational structure.
- The personality of the PM.
- The constraints under which they are working.

Try to specify a generic list of characteristics for a successful PM would be impossible. The list of characteristics that is presented below should be treated therefore as scales on which:

- Initially projects should be rated to determine the requirements of the person for the job.
- Individuals can be rated to determine their suitability for the job in hand.

The characteristics of individuals may be classified as either attitudes or skills. Attitude are those characteristics which are determined by the way a person thinks about a particular issue and which are reflected in the intent of that person. The skills are those characteristics which a person has been trained in or has. Learned and determined the tools available to that person regarding "how to handle a particular situation".

Attitudes which are desirable for PM are:

- A desire not just to satisfy but to delight customers and stake holders alike ;

- Accepting of both challenges and responsibility.
- Being focused on action, rather than procrastination getting the job done rather than avoiding critical or difficult issues;
- A desire to make the best use of all resources minimise naste in all activities.
- Does not lose right of the light at the end of the tunnel is goal focused;
- Has personal integrity people find it very difficult to respect and take the authority of a person who has low integrity;
- Is flexible about the route that must be taken to achieve the stated end-goals;
- Has personal goals that are consistent with those of the project organisation. The project team perceives that the pm and the organisation are going the same way.

Skills which are desirable include:

- Ability to determine the real needs.
- Desires of the customer this is done through 'getting close' to the customer via visits and both formal and informal discussions, and asking the relevant questions.
- Analytical skills to turn data into information and break down the project into comprehensible components parts;
- Technical skills–the PM need not be a technical specialist, but must at least be capable of comprehending the work that is being carried out and 'speaking the language' of the people involved.
- Team skills many battles have been won against poor odds by the ability of individuals to motivate and enthuse a team.
- Ability to delegate effectively not trying to do everything personally.
- Ability to manage your own time you cannot expect to manage other people unless you can show that you can manage yourself.
- The balancing of stockholder perceptions of project progress (otherwise known as being able to 'sell ideas').
- Negotiating skill. Resolve potential conflict situations so that all parties can be said to have 'won'.
- Problem solving/facilitating problem solving.
- Questions all assumptions made by stakeholder at all stages of activities. And if you can keep head while all about you are losing theirs.... . The above are all internal factors of the person. Their role and the way that it is managed by the organisation are just as important. This consideration will centre around four factors.

- **Responsibility.** The scope of the control of the PM as they perceive it.
- **Authority.** The organisation's view as to the scope of control of the PM.
- **Accountability.** The set of outcomes that are determined to be under the control of the PM.
- **Credibility.** The organisation must stand behind the PM and back the decisions that are made not doing so removes the individual's credibility and hence authority.

The rules for balancing these roles are:

- No responsibility without authority do not given a person the apparent responsibility for an area without giving them the authority to carry out their responsibilities.
- No authority without accountability under some can be held to account for their decisions or actions, they should not have the authority to make those decisions.

One of the most common causes of stress among PM is that they have apparent responsibility for an area, but not the authority to make the necessary changes in order for them to meet the requirements placed on them. This is termed 'role ambiguity'. It is common in organisation that are either only partly structured on project lines, or for whom the carrying out of project management in any formalised way is relatively novel. There are other stresses that be fall "pm" most notably due to need constantly to meet deadlines and the ever-changing nature of the work.

9 INTRODUCTION TO ENTERPRISE RESOURCE PLANNING (ERP)

LEARNING OBJECTIVES

- Introduction
- What is ERP?
- Business Modelling for ERP
- Salient Features for ERP Implementation
- Steps in Implementing ERP

INTRODUCTION

Globalization and Technological changes are the two forces which are influencing the national economical across the global. The global marketing is witnessing a shifting of power from the manufacturer to the giant retailers who are orienting themselves to customer needs. The organizations are under tremendous pressure to out perform their competitors. To retain their market share and competitive position, organizations are responding to unprecedented market demands and ever demanding customers by being cost competitive and now by reengineering the business process. This is changing the way the business is being done. Now, entering into Internet era, the core aspect of process focus in business and implementation of these aspects to achieve competitive advantage can be achieved through "Enterprise Resource Planning".

WHAT IS ERP?

Enterprise Resource Planning is defined as the process that is capable of planning all the resources of the enterprise to achieve a specified goal over a fixed period of time through system integration. ERP integrates all the functions and departments of an enterprise from suppliers to the end users. Every aspect of enterprise finds its place in ERP.

BUSINESS MODELLING FOR ERP

The pre-step to apply ERP approach to any business is the process mapping of the business. The development of a business map comprises of core business or the activities that are carried out by the business. It is a diagrammatic representation of the business as a main system indicating the sequence and interconnection of all the subsystems that the system comprises.

ERP is developed for a particular enterprise, based upon the long range plans of the business, business objectives and business modelling. ERP is developed in order to provide logistic support to the organization to manage the processes that are the part of the business model.

SALIENT FEATURES OF ERP IMPLEMENTATION

ERP brings together people with different functions and it produces, develops and leads to a sweeping change throughout the organization. Thus, the implementation of ERP should be managed by a proper change strategy. ERP demands the commitment of time and resources and also the best people to handle it to ensure the success. The role of a consultant is crucial in implementing ERP and the consultant should be able to set the expectations of the users at various levels keeping in mind the overall objectives of the business.

Customization of ERP is essential to make it suitable by changing the core package. The ERP should be customized to suit the business objections and user requirements.

The important issues which contribute to the success of ERP are:

1. Functionality
2. Technology
3. Implementability of the solution.

STEPS IN IMPLEMENTING ERP

The various steps involved in implementation of ERP are:

- Identification of the needs of the ERP implementation package.
  - Need for quick flow of information
  - Effective management information system for quick decision making
  - Elimination of manual records handling
  - High level of integration between functions.
- Evaluation of existing situation of business.
  - Thorough understanding of existing process
  - Procedures and methods for performing processes
  - Evaluation criteria for assessment
  - Present records, documents and manuals.
- Decision regarding the desired situation of business through bench marking.
- Re-engineering the business processes to achieve the desired results.
  - Reduce cycle time
  - Reduction of number of decision points
  - Streamlining the flow of information.
- Evaluation of various ERP packages. The available ERP packages are to be evaluated on the basis of:
  - Local or Global presence
  - Target market
  - Modularity
  - Cost of implementation
  - Post-implementation support
  - Investment in R&D
  - Price
  - Obsolescence
  - Ease of implementation.
- Finalization of ERP package
- Installations of necessary hardware and networks
- Finalization of ERP consultant for implementation
- Implementation of ERP in phased manner.

Guidelines for ERP implementation are given as follows:

Guidelines for ERP Implementation

- Understand your corporate culture and needs.
- Make the necessary business process changes.
- Communicate across the organization.
- Provide strong leadership.
- Chose the balanced team and suitable methodology.
- Train everyone to gain acceptability and commitment.

The steps implementations of ERP are represented as follows:

1. Identification of needs of ERP implementation package
2. Evaluate existing business situation
3. Decision regarding desired situation
4. Re-engineer business process
5. Evaluate various available ERP packages
6. Finalize the one best package, that suits your requirement
7. Install necessary hardware
8. Finalize the ERP consultant
9. Implement a phased manner

Steps in implementing ERP

TOTAL QUALITY MANAGEMENT (TQM)

Total Quality Management is concerned with the integration of all the efforts in the organization towards quality improvements, development and maintenance to meet full customer satisfaction at minimum cost.

Recent years have been revolutions in the manner in which organization approach quality. Quality is no longer seen as something that can be added "on" or inspected at the end of the process. A more systematic approach is necessary to build-in quality which calls for uncompromising commitment to the quality, from the top management to down, throughout the organization.

TQM means a systematic approach which integrates quality related activities throughout the organization including market research, research and development, operational planning for production of goods and services, procurement and production or services. Taken together, all these activities produce goods and services economically, meet the requirements of customers and are planned to achieve key organizational goals set by top management relative to quality. Total quality efforts enhance quality of work life, employee satisfaction through participation and involvement and consequently the image of the organization.

TQM model is shown in Fig. 9.1. This model distinctly different way of looking at the management style. It develops a participative culture where each employee can directly participate in decision making concerning his work. It is organized through quality circles, on a voluntary basis and quality improvement teams. It builds positive attitude towards quality and organization and enhances respect for each other leading to a meaningful workplace proud to be in.

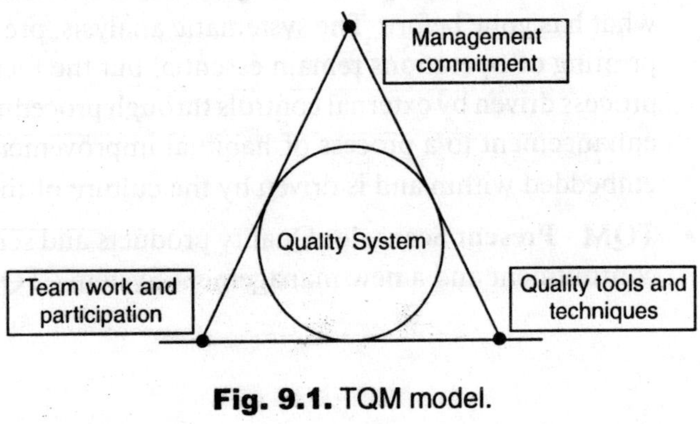

Fig. 9.1. TQM model.

EVOLUTION OF TQM

Morris Foster and Susan Whittle in their paper "the quality management maze" reviewed the development of TQM. To avoid the confusion arising out of ill defined and inconsistent terms of Quality Control and Quality Management.

The quality management can be classified into four phases:

- **Quality Control (QC) Phase:** Quality control focuses on product defect detention through post-production inspection. It is concerned with adherence to standards and sorting our rejects.
- **Quality Assurance (QA) Phase:** Quality assurance recognizes that "inspection is not an answer" and that the entire manufacturing process must be committed to meeting the quality needs of the design. QA focuses on procedure compliance and product conformity to specifications through production/operation management often using statistical process control techniques.
- **Total Quality Control (TQC) Phase:** TQC programmes attempts to expand the QA philosophy beyond manufacturing operations on other areas of organizational life. TQC incorporates many of the tools used in Quality Assurance but the purpose of problem analysis is to develop long term solutions rather than respond to short term variations. It is concerned with direct cost reduction and pre-elimination of waste and non-value adding procedures. TQC assures continuous improvement through refinement and expansion of quality control systems and procedures TQC is expected to be companywide movement but actually is largely limited to the manufacturing department.
- **Total Quality Management (TQM):** TQM is a fundamental shift from what has gone before. The systematic analysis, pre-planning and blue printing of operations remain essential but the focus switches from a process driven by external controls through procedure compliance and enhancement to a process of habitual improvement where control is embedded within and is driven by the culture of the organization.
- **TQM - Present Scenario:** Quality products and services demand and commitment and a new management System - TQM

Dr. Steve Smith points out that the following results are almost invariable through TQM.

1. Committed customers
2. Improved productivity
3. Reduced costs
4. Improved certainty in operations
5. Improved company image
6. Dedicated management
7. Increased employee participation.

PRINCIPLES OF TQM

Total Quality Management is an integrated management concept for continuous improvement of quality of goods/services through the participation at all the levels and functions of the organization. It is a process of building quality into goods and services from the start and making quality everyone's concern and responsibility. The success of TQM depends on the genuine commitment to quality of every member of the organization. TQM incorporates several dimensions.

(*a*) Design of products that meet customer's needs

(*b*) Control of process of ensure their ability to meet design requirements and needs

(*c*) Quality improvement for the continued enhancement of quality.

Customer is the driving force behind quality of design. The assurance of customer quality satisfaction must begin during new product development.

TOOLS AND TECHNIQUES OF TQM

Techniques for design control are

(*a*) Quality Function Deployment (QFD)

(*b*) Design review

(*c*) Statistical design of experiments

(*d*) Failure Mode-Effect and Critically Analysis (FMECA)

- ➢ Process control techniques are:
  - (*a*) Control charts
  - (*b*) Acceptance sampling
  - (*c*) Vendor rating.
- ➢ Techniques for implementing quality programme
  - (*a*) Daily process management
  - (*b*) Use of manuals and instructions
  - (*c*) First piece inspection.
- ➢ Principles of TQM can be stated as
  - (*a*) Customer focus
  - (*b*) Management led
  - (*c*) Company wide
  - (*d*) Do right first time, all the time
  - (*e*) Everyone's responsibility
  - (*f*) Continuous improvement
  - (*g*) Prevention, not detection
  - (*h*) Empower people
  - (*i*) Cost of quality.

IMPLEMENTATION OF TQM PROGRAMME

Successful implementation of TQM depends upon several key concepts. The key concepts are:

1. **Long Term Perspective:** Improvements do not occur overnight. The planning and organizing of improvement activities take time and require major commitments from each one in the organization.
2. **Customer Focus:** This is based on the premises that the quality is driven and defined by customer.
3. **Top Management Commitment:** Absence of Commitment of top managers will lead to lack of initiative and lower the spirit. Quality should be a major factor in strategic planning and competitive analysis.
4. **Systems Thinking:** TQM is effective only when all the employees are part of TQM implementation. TQM requires horizontal integration between organizational units. A process focus, by which inputs are transformed into outputs provides better insight into how organization actually operates.

5. **Training:** Each one in organization is required to be trained in TQM philosophies and techniques.
6. **Participation:** Everyone must participate in improvement efforts.
7. **Measurement and Reporting System:** Reporting of information must be timely, accurate and systematic process to measure and evaluate quality continuously is necessary.
8. **Communication:** TQM requires improved communication to facilitate improvements. People need to communicate across organizational levels, functions, product lines and locations to solve problems and implement change.
9. **Strong Leadership:** The leader must be determined to establish TQM initiatives and committed to sustaining its activities through daily actions.

BENEFITS OF TQM

- Enhancement of quality of goods and services
- Improving customer satisfaction
- Cost reduction and improvement in productivity
- Flexibility to accommodate to meet market demands
- Enhanced Quality of Work LIFE (QWL)
- Better utilization of human resources and empowerment of people.

PROBES IN IMPLEMENTATION OF TQM

The main problems in implementation of TQM programmed arises because of two facts;

1. **TQM is a Change:** TQM requires significant change in the organization - change in methods, processes, attitude of people and behaviour. It takes time to accept this change.
2. **Labour Management Relation:** The success of the TQM programme depends upon the relationship between labour and management. The participation of people at all levels is a prerequisite for TQM programme implementation.

ISO-9000 QUALITY STANDARDS

International Organization for Standards (ISO) introduced ISO-9000 series of quality standards in 1987 as a asset of guidelines for developing and demonstrating a documenting a documented quality system. This is to facilitate international trade of goods and services. The ISO-9000 quality standards have revolutionized quality system management in recent years. Current trends show ISO-9000 will play a vital role in unifying international community in bilateral and multinational cooperation.

ISO-9000 is a series of five international standards for quality management and quality assurance system which are applicable to all types of organizations irrespective of the nature, size and background. ISO-9000 series define the discipline and specify the procedures and criteria to ensure that the output of the organization consistently meets the customers requirements. ISO standards guide to establish, operate, document and maintain an effective, economic quality system which will demonstrate to the customer as well as to the management that organization is committed to quality and is capable of meeting the specified requirement.

OBJECTIVES

The basic objectives of ISO-9000 are:

- TO facilitate international trade of goods and services
- To obtain competitiveness by obtaining required quality in a cost effective way
- To develop mutual cooperation in respect of intellectual scientific, technological and economic activities.

STRUCTURE OF ISO-9000 STANDARDS

ISO-9000 series is a set of five international standards and supplemented by a vocabulary standard defining the various terms used in the series. These standards are shown in Fig. 9.2.

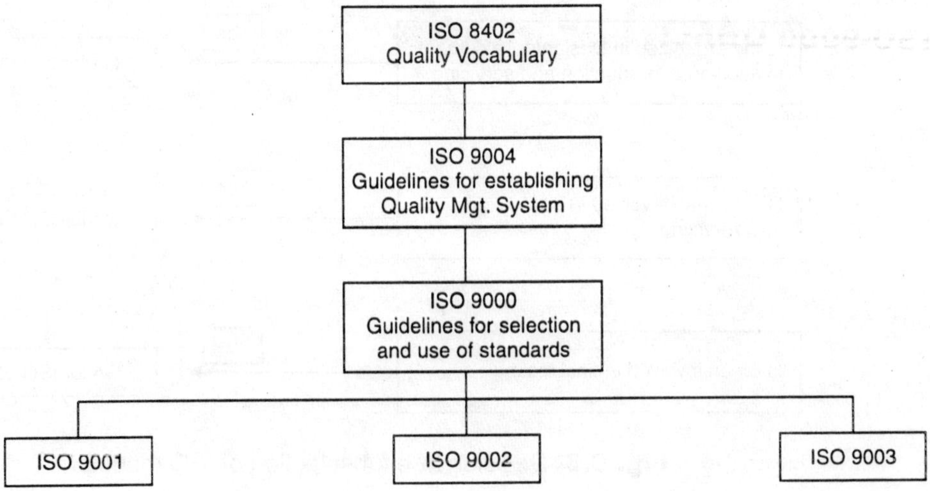

Fig. 9.2. Structure of ISO 9000 Standards.

SELECTION AND USE OF STANDARDS

An organization which intends to adopt ISO-9000 quality management system has to select one of the three quality system models—ISO-9001, ISO-9002, or ISO-9003.

- **ISO-9001 System:** Model for quality assurance in design/development, production installation and serving for use when conformance to specification is to be assured by the supplier during production and installation.
- **ISO-9002 System:** Model for quality assurance in production, installation and servicing. For use when conformance to specified requirements is to be assured by the supplier during production and servicing.
- **ISO-9003 System:** Model for quality assurance in final inspection and testing. For use when conformance to required specification to be assured by the supplier solely to final inspection. Fig. 9.3 shows the decision tree for selection of models.

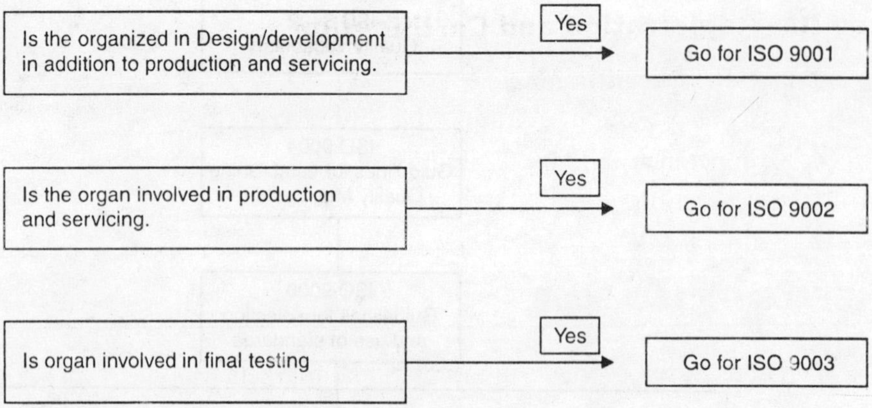

Fig. 9.3. Decision tree for selection of ISO models.

IMPLEMENTATION OF ISO-9000 QUALITY MANAGEMENT SYSTEM

The various phases in the installation of quality management system are:

Phase 1—Preparation phase

1. Management commitment to ISO-9000 quality management system
2. Setting up of implementation committee
3. Quality awareness programme
4. Data collection of existing system status and preparation plan
5. Develop organization structure
6. Develop the documentation system
7. Develop resource system.

Phase II—Implementation

8. Implement documented quality system
9. Establish internal quality audit system
10. Monitor and stabilize the quality system
11. Conduct pre-registration internal audit.

Phase III—Registration and Certification

12. Apply for registration
13. Adequacy audit by certifying agency
14. Competence audit by certifying body
15. Certification.

BENEFITS OF ISO-9000

1. Helps to sell products in international markets
2. Sets the base line and brings about discipline in systems, procedures and documentation
3. Improves quality awareness
4. Improves the confidence of customer in company's product and hence the company image is enhanced.

Review Questions

1. What is Enterprise Resource Planning (ERP). Explain briefly.
2. How is ERP applied in business modelling?
3. What are the steps in ERP implementation? Write the salient feature for ERP implementation.
4. Define total quality management (TQM). Which steps are involved in implementation of TQM programme?
5. Define structure and uses of ISO-9000 standards.
6. What are the various phases in the installation of quality management system?
7. Write benefits of ISO-9000.

10 NETWORK ANALYSIS

LEARNING OBJECTIVES

- Introduction
- What is Network Analysis?
- Business Modelling for Network Analysis
- Salient Features for Network Analysis
- Meaning of PERT and CPM
- Network Diagram Construction
- 20 Key Project Manager Actions and Results
- 14 Key Principles for Project Management Success
- Materials Requirements Planning (MRP)

NETWORK ANALYSIS—AN INTRODUCTION

Network analysis is the general name given to certain specific techniques which can be used for the planning, management and control of projects. According to Project Management Institute, the definition of a project is

A project is a temporary endeavour undertaken to create a "unique" product or service.

This definition serves to highlight some essential features of a project, such as:

- It is temporary - it has a beginning and an end
- It is "unique" in some way.

With regard to the use of the word unique, we personally prefer to use the idea of "non-repetitive" or "non-routine", *e.g.,* building the very first Boeing Jumbo jet was a project-building than now is a repetitive/routine manufacturing process, not a project.

We can think of many projects in real-life, *e.g.,* building the Channel tunnel, building the London Eye, developing a new drug, etc.

Typically, all projects can be broken down into:

- **separate activities (tasks/jobs)** - where each activity has an associated duration or *completion time* (*i.e.,* the time from the start of the activity to its finish)
- **precedence relationships** - which govern the order in which we may perform the activities, *e.g.,* in a project concerned with building a house the activity "erect all four walls" must be finished before the activity "put roof on" can start and the problem is to bring all these activities together in a coherent fashion to complete the project.

Two different techniques for network analysis were developed independently in the late 1950's, these were:

- PERT (for Program Evaluation and Review Technique); and
- CPM (for Critical Path Management).

PERT

PERT was developed to aid the US Navy in the planning and control of its Polaris missile program. This was a project to build a strategic weapons system, namely the first submarine launched intercontinental ballistic missile, at the time of the Cold War between the USA and Russia. Military doctrine at that time emphasized 'MAD—Mutually Assured Destruction', namely if the other side struck first then sufficient nuclear weapons would remain to obliterate their homeland. That way peace was preserved. By the late 1950s the USA believed (or more importantly believed that the Russians believed) that American land based missiles and nuclear bombers were vulnerable to a first strike. Hence there was a strategic emphasis on completing the Polaris project as quickly as possible, cost was not an issue. However no one had ever built a submarine launched intercontinental ballistic missile before, so dealing with uncertainty was a key issue. PERT has the ability to cope with uncertain activity completion times (*e.g.*, for a particular activity the most likely completion time is 4 weeks but it could be anytime between 3 weeks and 8 weeks).

CPM

CPM was developed in the 1950's as a result of a joint effort by the Dupont Company and Remington Rand Univac. As these were commercial companies cost was an issue, unlike the Polaris project mentioned above. In CPM the emphasis is on the trade-off between the cost of the project and its overall completion time (*e.g.*, for certain activities it may be possible to decrease their completion times by spending more money - how does this affect the overall completion time of the project?)

Modern commercial software packages tend to blur the distinction between PERT and CPM and include options for uncertain activity completion times and project completion time/project cost trade-off analysis. Note here that many such packages exist for doing network analysis.

There is no clear terminology in the literature and you will see this area referred to by the phrases: network analysis, PERT, CPM, PERT/CPM, critical path analysis and project planning.

Network analysis is a vital technique in PROJECT MANAGEMENT. It enables us to take a *systematic quantitative structured approach* to the problem of managing a project through to successful completion. Moreover, as will become clear below, it has a graphical representation which means it can be understood and used by those with a less technical background.

Example

We will illustrate network analysis with reference to the following example: suppose that we are going to carry out a minor redesign of a product and its associated packaging. We intend to test market this redesigned product and then revise it in the light of the test market results, finally presenting the results to the Board of the company.

The key question is: *How long will it take to complete this project?*

After much thought we have identified the following list of separate activities together with their associated completion times.

Activity Number		Completion Time (Weeks)
1.	Redesign product	6
2.	Redesign packaging	2
3.	Order and receive components for redesigned product	3
4.	Order and receive material for redesigned packaging	2
5.	Assemble products	4
6.	Make up packaging	1
7.	Package redesigned product	1
8.	Test market redesigned product	6
9.	Revise redesigned product	3
10.	Revise redesigned packaging	1
11.	Present results to the Board	1

It is clear that in constructing this list of activities, we must make judgments as to the level of detail (timescale) to adopt. At one extreme we could have just a single activity "do the projects" and at the other extreme we could try to break the project down into hourly activities. The appropriate timescale to adopt (which can be different for different activities) grows out of our knowledge of the situation and experience.

Aside from this list of activities we must also prepare a list of precedence relationships indicating activities which, because of the logic of the situation, must be finished before other activities can start *e.g.,* in the above list activity number 1 must be finished before activity number 3 can start.

It is important to note that, for clarity, we try to keep this list to a minimum by specifying only *immediate* relationships, which are relationships involving activities that "occur near to each other in time".

For example it is plain that activity 1 must be finished before activity 9 can start but these two activities can hardly be said to have an immediate relationship (since many other activities after activity 1 need to be finished before we can start activity 9).

Activities 8 and 9 would be examples of activities that have an immediate relationship (activity 8 must be finished before activity 9 can start).

Note here that specifying non-immediate relationships merely complicates the calculations that need to be done - it does not affect the final result. Note too that, in the real-world, the consequences of missing out precedence relationships are much more serious than the consequences of including unnecessary (non-immediate) relationships.

Again after much thought (and aided by the fact that we listed the activities in a logical/chronological order) we come up with the following list of immediate precedence relationships.

Activity number		Activity number	
1	must be finished before	3	can start
2		4	
3		5	
4		6	
5, 6		7	
7		8	
8		9	
8		10	
9, 10		11	

The key to constructing this table is, for each activity in turn, to ask the question:

"What activities must be finished before this activity can start?"

Note here that:

- activities 1 and 2 do not appear in the right hand column of the above table, this is because there are no activities which must finish before they can start, *i.e.,* both activities 1 and 2 can start immediately
- two activities (5 and 6) must be finished before activity 7 can start
- It is plain from this table that non-immediate precedence relationships (*e.g.,* "activity 1 must be finished before activity 9 can start") need not be included in the list since they can be deduced from the relationships already in the list.

Once we have completed our list of activities and our list of precedence relationships we combine them into a diagram/picture (called a *network* - which is where the name network analysis comes from). We do this below.

Note first however that we asked the key question above:

How long will it take to complete this project?

(*i.e.*, complete all the activities whilst respecting the precedence relationships). What would you say? *e.g.,*

- Could we complete this project in 3 years?
- Could we complete this project in 2 weeks?

One answer could be if we first do activity 1, then activity 2, then activity 3, then activity 10, then activity 11. Such an arrangement would be possible here, (check the precedence relationships above), and the project would then take the sum of the activity completion times, 30 weeks.

However could we complete the project in less time? It is clear that logically we need to amend our key question to be:

What is the minimum possible time in which we can complete this project?

We shall see below how the network analysis diagram/picture we construct helps us to answer this question.

Note too here that, at this stage, we are merely planning the project - we have not started any of the various activities mentioned above.

Network diagram construction

The network diagram we construct is different depending upon whether we are going to use an activity on node or an activity on arc network.

As will become apparent below this network diagram represents clearly and concisely the entire project - enabling us to get a pictorial overview of what is to happen, the various activities and their relationship one to another.

Now go to activity on node or activity on arc. If you are new to this topic then I suggest you use activity on node as it is (at least in my view) easier to follow. Also note here that (to prevent duplication of material) it is only under activity on note that I discuss the computer (package) solution of the above example.

Project Management Institute

Project Management Institute (PMI), with more than 170,000 members in over 150 countries, is the world's foremost advocate for the project management profession. A vital and forward-thinking organization, PMI is comprised of 247 chartered chapters, 16 new potential chapters, 30 Specific Interests Groups (SIGs) and two colleges. PMI sets industry standards, conducts research and provides education, certification and professional exchange opportunities designed to strengthen and further establish the profession. PMI advances the careers of practitioners, while enhancing overall business and government performance, through documentation of return on investment.

Project management is a strategic competency that has successfully been applied in such high profile projects as rebuilding the Pentagon after the 9/11 terrorist attacks, the transformation of IBM by Lou Gerstner from an IT hardware giant to an IT software and computer services leader, organizing and managing the Salt Lake City Winter Olympics, and the upcoming Summer Games in Beijing, just to name a few. If project management can play a major role in these success stories, just imagine what it might be able to do for your own organization.

20 KEY PROJECT MANAGER ACTIONS AND RESULTS

The Project Manager s Partner by Michael Greer is organized around these 20 Key Project Manager Actions and Results. The handbook provides several pages of background information concerning each of the actions, as well as step-by-step guidelines, worksheets, and examples for helping new project managers achieve results quickly and easily. The 20 Key Project Manager Actions are organized according to their support of the Five Essential Project Management Processes: Initiating, Planning, Executing, Controlling, and Closing.

Action	Results of Successful Performance
	Initiating
1. Demonstrate project need and feasibility.	A document confirming that there is a need for the project deliverables and describing, in broad terms: the deliverables, means of creating the deliverables, costs of creating and implementing the deliverables, benefits to be obtained by implementing the deliverables.
2. Obtain project authorization.	A "go/no go" decision is made by the sponsor. A project manager is assigned. A "project charter" is created which: Formally recognizes the project Is issued by a manager external to the project and at a high enough organizational level so that he or she can meet project needs Authorizes the project manager to apply resources to project activities.
3. Obtain authorization for the phase.	A "go/no go" decision is made by the sponsor which authorizes the project manager to apply organizational resources to the activities of a particular phase Written approval of the phase is created which: Formally recognizes the existence of the phase Is issued by a manager external to the project and at a high enough organizational level so that he or she can meet project needs.

	Planning
4. Describe project scope.	Statement of project scope Scope management plan Work break down structure.
5. Define and sequence project activities.	An activity list (list of all activities that will be performed on the project) Updates to the Work Break down Structure (WBS).
	A project network diagram
6. Estimate durations for activities and resources required.	Estimate of durations (time required) for each activity and assumptions related to each estimate Statement of resource requirements Updates to activity list.
7. Develop a project schedule.	Project schedule in the form of Gantt charts, network diagrams, milestone charts, or text tables Supporting details, such as resource usage overtime, cash flow projections, order/delivery schedules, etc.
8. Estimate costs.	Cost estimates for completing each activity Supporting detail, including assumptions and constraints cost management plan describing how cost variances will be handled.
9. Build a budget and spending plan.	A cost baseline or time-phased budget for measuring/ monitoring costs A spending plan, telling how much will be spent on what resources at what time.
10. Create a formal quality plan. (*optional*)	Quality management plan, including operational definitions Quality verification checklists.
11. Create a formal project communications plan. (*optional*)	A communication management plan, including: Collection structure Distribution structure Description of information to be disseminated Schedules listing when information will be produced A method for updating the communications plan.

12. Organize and acquire staff.	Role and responsibility assignments Staffing plan Organizational chart with detail as appropriate Project staff Project team directory.
13. Identify risks and plan to respond. (*optional*)	A document describing potential risks, including their sources, symptoms, and ways to address them.
14. Plan for and acquire outside resources. (*optional*)	Procurement management plan describing how contractors will be obtained Statement of Work (SOW) or Statement of Requirements (SOR) describing the item (product or service) to be procured Bid documents, such as RFP (Request For Proposal), IFB (Invitation For Bid), etc. Evaluation criteria—means of scoring contractor's proposals Contract with one or more suppliers of goods or services.
15. Organize the project plan.	A comprehensive project plan that pulls together all the outputs of the preceding project planning activities
16. Close out the project planning phase.	A project plan that has been approved, in writing, by the sponsor A "green light" or okay to begin work on the project.
17. Revisit the project plan and replan if needed.	Confidence that the detailed plans to execute a particular phase are still accurate and will effectively achieve results as planned.
Executing	
18. Execute project activities.	Work results (deliverables) are created Change requests (*i.e.,* based on expanded or contracted project) are identified Periodic progress reports are created Team performance is assessed, guided, and improved if needed Bids/proposals for deliverables are solicited, contractors (suppliers) are chosen, and contracts are established Contracts are administered to achieve desired work results.

Controlling	
19. Control project activities.	Decision to accept inspected deliverables Corrective actions such as rework of deliverables, adjustments Updates to project plan and scope List of lessons learned Improved quality Completed evaluation check lists (if applicable).
Closing	
20. Close out prjoect activities.	Formal acceptance, documented in writing, that the sponsor has accepted the product of this phase or activity Formal acceptance of contractor work products and updates to the contractor's files Updated project records prepared for archiving A plan for follow-up and/or hand-off of work products

Materials Requirements Planning (MRP)

Materials requirements planning, referred to the initials MRP, is a technique which assists a company in the detailed planning of its production. Recall here that the master production schedule sets out an aggregate plan for production. MRP translates that aggregate plan into an extremely detailed plan. We shall introduce MRP by means of an example.

Example

The production manager at Alders Hot manufacturing wishes to develop a materials requirements plan for producing chairs over an 8 week period. She estimates that the lead time between releasing an order to the shop floor and producing a finished chair is 2 weeks. The company currently has 260 chairs in stock and no safety stock (safety stock is stock held in reserve to meet customer demand if necessary). The forecast customer demand is for 150 chairs in week 170 in week 3, 175 in week 5, 90 in week 7 and 60 in week 8.

It helps to understand that what is going on if we write out, overtime, the demand for chairs as below.

Week	1	2	3	4	5	6	7	8
Demand	150	0	70	0	175	0	90	60
On-hand at end of week 110								
Order	?	?	?	?	?	?	?	?

Here, we have shown the demand in each of the eight weeks, initially we have 260 chairs available so if these are used to meet the demand of 150 in week 1 we have 260 – 150 = 110 left on-hand (*i.e.*, in stock) at the end of the week. Plainly we will need to order some more chairs in order to meet all of the forecast future demand over the 8 week planning period.

Conceptually therefore we face two related decisions about ordering:

- **timing** - when to order
- **quantity** - how much to order

You can think of asking yourself the question, in each and every period, *should I order in this period and if so how much?*

For the moment, suppose we order nothing in week 1, nothing in week 2, etc. The situation by the time we reach the end of week 5 will be as below:

Week	1	2	3	4	5	6	7	8
Demand	150	0	70	0	175	0	90	60
On-hand at end of week	110	110	40	40	- 135			
Order	?	?	?	?	?	?	?	?

If we are to avoid a stock out in week 5 we plainly need to order at least 135 chairs. Now we know that the lead time between ordering a chair and receiving it is 2 weeks. Therefore to avoid a stock out in week 5 we must have ordered 135 chairs either in week 3, or in any week before week 3. In other words ordering

- 135 chairs in week 1, or
- 135 chairs in week 2, or
- 135 chairs in week 3,

Would each ensure that we have sufficient chairs available to meet forecast demand in week 5.

If we order these chairs earlier than week 3 we will be carrying extra inventory (stock) for a number of periods and, as we know, carrying stock costs money.

It would seem appropriate therefore to order 135 chairs in week 3. This will give:

Week	1	2	3	4	5	6	7	8
Demand	150	0	70	0	175	0	90	60
On-hand at end of week	110	10	40	40	0			
Order	0	0	135	?	?	?	?	?

Continuing on in the same manner we get:

Week	1	2	3	4	5	6	7	8
Demand	150	0	70	0	175	0	90	60
On-hand at end of week	110	110	40	40	0	0	– 90	
Order	0	0	135	?	?	?	?	?

requiring an order of 90 chairs in week 5 and giving:

Week	1	2	3	4	5	6	7	8
Demand	150	0	70	0	175	0	90	60
On-hand at end of week	110	110	40	40	0	0	0	
Order	0	0	135	0	90	?	?	?

Continuing again we get:

Week	1	2	3	4	5	6	7	8
Demand	150	0	70	0	175	0	90	60
On-hand at end of week	110	110	40	40	0	0	0	–60
Order	0	0	135	0	90	?	?	?

requiring an order of 60 chairs in week 6 and giving:

Week	1	2	3	4	5	6	7	8
Demand	150	0	70	0	175	0	90	60
On-hand at end of week	110	110	40	40	0	0	0	0
Order	0	0	135	0	90	60	?	?

Note that we have no data given here on which to base order decisions in weeks 7 and 8. As we are at the end of the planning period these are usually taken as zero.

Decisions

Let us be clear about what we have done here with respect to our two decisions of:

- timing - when to order
- quantity - how much to order

With respect to the timing decision we always ordered *as late as possible*, but never planned a stockout. This is a driving principle in MRP, never order before you need to, never plan to stockout.

With respect to the quantity decision we always ordered as little as possible, *i.e.* just enough to avoid a stockout. This is known as the lot for lot rule, sometimes called LFL or L4L or LL rule. This quantity decision rule can be varied in MRP and some other rules are:

- fixed order quantity rule (sometimes called FOQ or FO) - the quantity ordered is an integer multiple of the same fixed amount each time an order is made
- fixed period requirements rule (sometimes called FPR) - the quantity ordered should be enough for a fixed number of periods.

To illustrate the FPR rule suppose that we decide to order enough for 3 weeks when we make an order. The situation at the end of week 5 is (from above) repeated below:

Week	1	2	3	4	5	6	7	8
Demand	150	0	70	0	175	0	90	60
On-hand at end of week	110	110	40	40	– 135			
Order	?	?	?	?	?	?	?	?

To decide the FPR order quantity we continue this table until week 7. The quantity ordered must then be just sufficient to cover weeks 5 to 7 (*i.e.*, to cover 3 weeks as required for a 3 week FPR). This is done below:

Week	1	2	3	4	5	6	7	8
Demand	150	0	70	0	175	0	90	60
On-hand at end of week	110	110	40	40	–135	– 135	– 225	
Order	?	?	?	?	?	?	?	?

Hence the 3 week FPR order is 225 units in week 3, giving the situation below:

Week	1	2	3	4	5	6	7	8
Demand	150	0	70	0	175	0	90	60

On-hand at end of week	110	110	40	40	90	90	0	- 60
Order	0	0	225	?	?	?	?	?

Plainly we need some order in week 6 to cover the stockout in week 8. As we are at the end of the planning period we usually order just sufficient (*i.e.*, revert to the LFL rule) and order 60 in week 6 to give:

Week	1	2	3	4	5	6	7	8
Demand	150	0	70	0	175	0	90	60
On-hand at end of week	110	110	40	40	90	90	0	0
Order	0	0	225	0	0	60	?	?

Note here that with the FPR rule applied to cover p periods you make (atmost) one order every p periods (ignoring any order necessary at the end of the planning period).

Choice of Lot Size Rule

How then are we to choose between different lot size rules (LFL, FPR and FOQ). We could, of course, redo our calculations with different lot size rules (*e.g.* a FOQ ordering multiples of 100 each time). We would see their different effects but would still have to choose between them.

All of the rules ensure forecast demand is meet, *i.e.* no stockouts, so this is not a distinguishing feature. The LFL rule (by ordering as little as possible each time) will keep average inventory levels low, but will result in more orders on average. Both the FPR and FOQ rules will have higher inventory levels, but will result in less orders on average. Choosing a lot size rule therefore comes down to balancing the number of orders against the cost of holding inventory, just as we considered in deriving the FOQ formula in the inventory notes. Hence, given cost information, it is possible to derive the most effective (least costly) lot rule to use for any particular item.

Extending the Example

Whilst for the example considered above, just a single item, we easily worked out the orders manually it is obvious that as the number of items increases, a manual calculation becomes too complicated and we need a computer package. We illustrate this below.

For the chair production problem considered before suppose now that the production manager as well as planning the production of the chair must also plan the production of the components that make up the chair. These are: the seat, a back and four legs. The lead time for seats and backs is 2 weeks and the lead time for legs is one week. The company currently has an inventory of 60 seats, 40 backs and 80 legs. Scheduled receipts are 50 seats in week 1 and 10 backs in week 1. The lot size rules the production manager has decided to adopt is LFL for for all items.

Now in planning the production of chairs we need also to plan the production of seats, backs and legs. For example we show below the situation as derived above where in week 3 we issued an order for 135 chairs.

Week	1	2	3	4	5	6	7	8
Demand	150	0	70	0	175	0	90	60
On-hand at end of week	110	110	40	40	0			
Order	0	0	135	?	?	?	?	?

Now to have 135 chairs made we need to have to hand (*i.e.* currently available) 135 seats, 135 backs and 4(135) = 540 legs. The current inventory of these items (plus scheduled receipts) is insufficient, so orders must be placed for these items. Just as we did for the chair itself above these orders must be *phased in time* so as to ensure that we never stock out.

Now to do all this manually for chairs, seats, backs and legs would just be too time consuming and error-prone. It would be far better to do this via as a computer package, such as the package used in this course.

In order to solve this example using the package we need to have a better overview of MRP.

MRP Overview

In MRP, two types of information are required:

- Structural; and
- Tactical.

Structural information is information about the items (parts/components) that the company uses and how different items are related to one another. It includes information for each item such as lead time and lot (or batch) size rule. The key point about this information is that it changes relatively infrequently.

Tactical information is information about the current state of the company for example sales orders (real and forecast) pending, the master production schedule, on-hand inventory levels and purchase orders. Obviously the key point about this information is that it changes frequently.

Structural Information

The structural information required in MRP relates to:

- Item information; and
- Bill of materials.

We deal with each of these in turn below.

Item Information

We have structural information relating to each of the items that we are producing. Below we give this structural information for each of the items in our simple example.

Item	*Lead time (weeks)*	*Lot size rule*	*Unit measure*	*ABC code*
chair	2	LFL	Each	A
seat	2	LFL	Each	B
back	2	LFL	Each	B
legs	1	LFL	Each	C

Note that:

- The lead time is the time between placing an order and receiving that order-note that this lead time assumes that all the items needed for production are available at the time at which we place an order. For example for the chair we, as mentioned previously, need one seat, one back and four legs to make it - the lead time of 2 weeks for a chair assumes that all of these items are available when we place an order for a chair - the 2 weeks covers the time required to assemble them into a chair.
- As mentioned above the lot size rule relates to deciding how large the lot (batch/order) should be each time an order is placed.
- In fact the package contains a total of ten different lot size rules.
- Unit measure is the units of measurement for each item - here it is each specifying that each item is identified individually.
- The ABC code relates to inventory analysis. Essentially items are divided into three inventory categories (A, B and C) where:
  - Category A items are the most important inventory items and total are responsible for 80% of inventory cost
  - Category B items are only moderately important inventory items and in total are responsible for 15% of inventory cost

- Category C items are relatively unimportant inventory items and in total are responsible for 5% of inventory cost.

Obviously the idea here is that management attention should be primarily focused upon category A items. In the package the ABC code is used purely to produce different management reports.

In turns of entering our example problem into the package the screen below is the initial screen we need to set up number of items, time units etc.

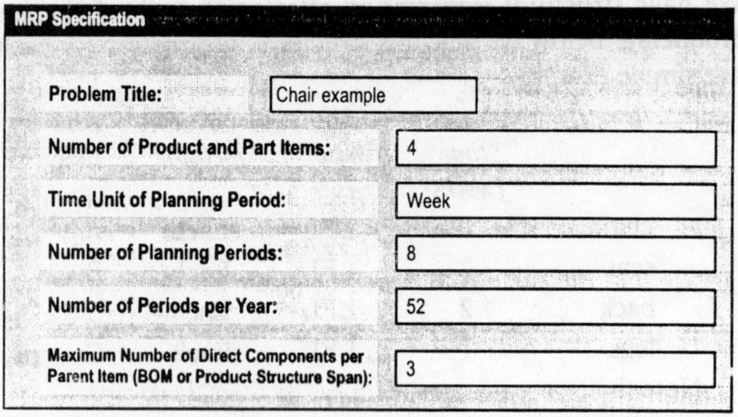

The package input for our items is as below

No.	Item ID	ABC Class	Source Code	Material Type	Unit Measure	Lead Time	Lot Szie
1	Chair	A			Each	2	LFL
2	Seat	B			Each	2	LFL
3	Back	B			Each	2	LFL
4	Legs	C			Each	1	LFL

Bill of Materials

In order to show the make-up (in terms of the parts needed for production) we have a *Bill of Materials (BOM)* for the end-product (namely the chair). Below we show the BOM for the chair.

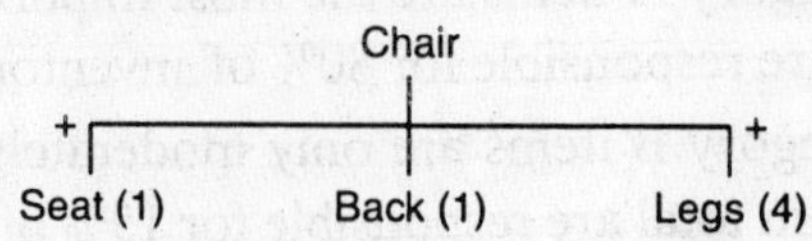

This BOM means that to produce one chair we need:

- one seat
- one back
- four legs

The Bill of Materials can be thought of as a *diagrammatic recipe*. Just as in cooking we need a list of ingredients and their quantities to know how to cook something, so here the BOM tells us what we need to make a chair.

The BOM is best thought of as being divided into levels, with the final item (the chair) being at the top level and the items needed to make up a chair being at the second level.

Other examples may have more levels, *e.g.*, if items at the second level are themselves made up from further items. Plainly BOM's are structural information that change relatively infrequently. It is also plain that any mistakes in specifying BOM's could have disastrous consequences on the shop floor - *e.g.*, consider what would happen if we fail to note that a particular part is needed in the production of some item.

In terms of the package we can enter this BOM as below.

Item ID	Component ID/Usage	Component ID/Usage	Component ID/Usage
Chair	Seat	Back	Legs/4
Seat			
Back			
Legs			

This BOM can also be shown graphically by the package, as below.

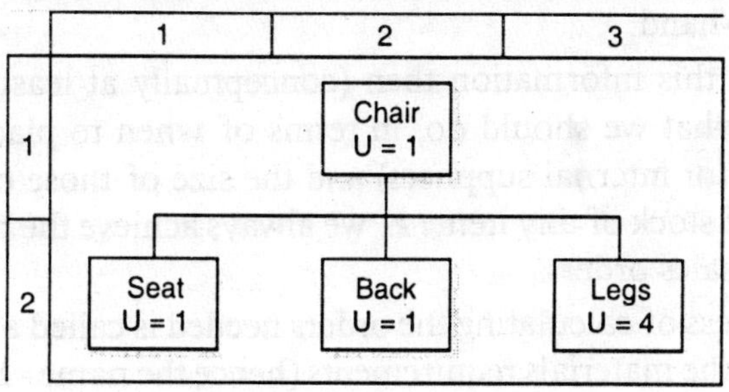

Tactical Information

The tactical information required in MRP relates to:

- Out-going inventory (sales) and planned production (master production schedule); and
- On-hand inventory and in-coming inventory (purchases).

We deal with each of these in turn below.

Diagrammatic Overview

Below we give a diagrammatic overview of the situation.

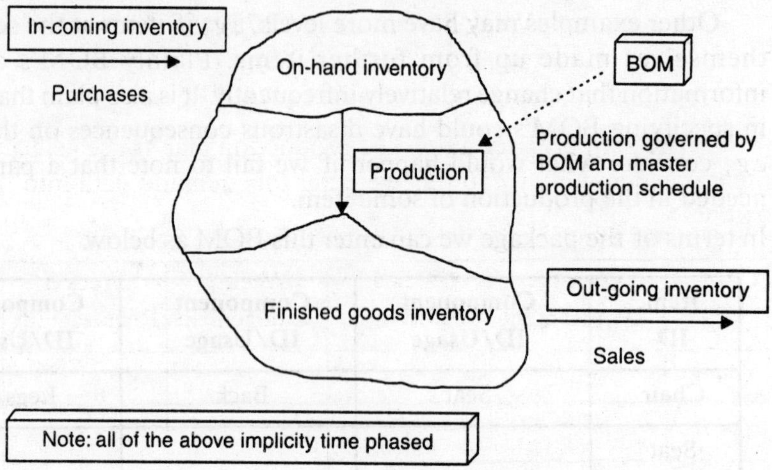

Fig. 10.1

We have specified above structural information about how the items relate to each other (BOM) and item information (*e.g.* lot size rule); and we will shortly (below) specify tactical information about planned production and inventory in/out/on-hand.

Given all this information then (conceptually at least) we should be able to *calculate* what we should do, in terms of when to place orders with external suppliers (or internal suppliers) and the size of those orders, so that we never run out of stock of any item *i.e.,* we always achieve the planned production and meet the sales orders.

This process of calculating the orders needed is called a *MRP EXPLOSION* and produces the materials requirements (hence the name - Materials Requirements Planning) .

An important point to note is that MRP is not cost driven *i.e.,* it does not seek to minimize cost (unlike the MPS system used in the package). Instead MRP is stock out driven - that is it will always order sufficient to avoid stock outs (using the lot size rule for each item) and order as late as possible. We can now specify the tactical information required.

Out-going Inventory and Planned Production

The table shows the current out-going inventory (predicted or expected sales). For our simple example this is just the forecast demand for chairs.

Time	*Quantity*
Week 1	150
Week 3	70
Week 5	175
Week 7	90
Week 8	60

We need to produce a detailed production plan for seats, backs and legs to meet this demand. To do this we enter this demand data into the package as below.

Item ID	Overdue Requirement	Week 1 Requirement	Week 2 Requirement	Week 3 Requirement	Week 4 Requirement	Week 5 Requirement	Week 6 Requirement	Week 7 Requirement	Week 8 Requirement
Chair		150		70		175		90	60
Seat									
Back									
Legs									

Then a detailed production plan can be obtained via an MRP explosion. We can however use the MRP package another way - to examine meeting a predetermined production schedule for chairs. Above we have just taken our chair requirements as relating to the total demand. We could however, have entered any numbers there - for example a chair production plan as determined from a cost based Master Production Schedule (MPS). In that case we simply enter the chair figures from the MPS as the chair requirements and the package will produce a detailed production plan for seats, backs and legs to met those requirements. For the simple example we consider here we shall just take the chair requirements as equal to the demand data.

The *feasibility* in terms of production capacity etc of producing the seats, backs and legs in any period to met our chair requirements is *not* considered in MRP. This is an important point. MRP typically takes no account of capacity considerations. Such considerations must be taken into account when the MPS is produced.

The package does not include the capability to take capacity into account. Although we can enter capacity data into the package this is only used for reporting proposes—to see whether we are exceeding production capacities or not. Capacities are not taken into account when the package decides what we need to produce and when.

On-hand and In-coming Inventory

On-hand inventory is the amount of inventory on-hand at the start of the planning period. For our simple example this is:

Item	*On-hand inventory (units)*
chair	260
seat	60
back	40
legs	80

In-coming inventory is the items we expect to receive as a result of buying-in (purchasing) from external suppliers, or items we expect to receive as a result of internal orders placed earlier. For our simple example this is:

Time	*Item*	*Quantity*
Week 1	seat	50
Week 1	back	10

This data is entered into the package as below.

Item ID	Safety Stock	On Hand Inventory	Overdue Planned Receipt	Week 1 Planned Receipt
Chair		260		
Seat		60		50
Back		40		10
Legs		80		

Output

Below we show annotated MRP output from the package (after explosion "Explode Materials Requirements") for the example given above for all items.

Chair

11-12-2000	**Overdue**	**Week 1**	**Week 2**	**Week 3**	**Week 4**	**Week 5**	**Week 6**	**Week 7**	**Week 8**	**Total**
Item: Chair		**LT = 2**	**SS = 0**	**LS = LFL**	**UM = Each**	**ABC = A**	**Source =**	**Type =**	**A. Demand = 0**	**@Cost = 0**
Gross Requirement	0	150	0	70	0	175	0	90	60	545
Scheduled Receipt	0	0	0	0	0	0	0	0	0	0
Projected On Hand	260	110	110	40	40	0	0	0	0	0
Projected Net Requirement	0	0	0	0		135	0	90	60	285
Planned Order Receipt	0	0	0	0	0	135	0	90	60	285
Planned Order Release	0	0	0	135	0	90	60	0	0	285

From the last line "Planned Order Release" we see that we need to order 135 chairs in week 3, 90 in week 5 and 60 in week 6. These are as we would expect, since for our simple example we manually calculated when we needed to order chairs to meet demand. The benefit of MRP is that the process is automated (hence quicker and less error-prone), and also that we can easily take other items that go to make up a chair (such as seat, back and legs) into account.

In the output above the columns represent different periods and the rows mean:

- ***Gross Requirement*** - the forecast demand for the item
- ***Scheduled Receipt*** - any scheduled receipts from previous orders that are currently being processed
- ***Projected On Hand*** - what the inventory level will be if we produce according to the plan above and the demand forecast is correct
- ***Projected Net Requirement*** - what we will need in each period to avoid a stock out, *i.e.*, the amount that is required to avoid stock outs
- ***Planned Order Receipt*** - what we will receive each period as a result of planned orders (the orders the package has automatically calculated are needed to avoid stock outs)
- ***Planned Order Release*** - when we need to release (make) the planned orders having regard to the lead time required

Seat

11-12-2000	**Overdue**	**Week 1**	**Week 2**	**Week 3**	**Week 4**
Item: Seat		**LT = 2**	**SS = 0**	**LS = LFL**	**UM = Each**
Gross Requirement	0	0	0	135	0
Scheduled Receipt	0	50	0	0	0
Projected On Hand	60	110	110	0	0
Projected Net Requirement	0	0	0	25	0
Planned Order Receipt	0	0	0	25	0
Planned Order Release	0	25	0	90	60

Week 5	Week 6	Week 7	Week 8	Total
ABC = B	Source =	Type =	A. Demand = 0	@Cost = 0
90	60	0	0	285
0	0	0	0	50
0	0	0	0	
90	60	0	0	175
90	60	0	0	175
0	0	0	0	175

From this output we can see that we are advised to place an order for 25 seats in week 1. This advice should be followed unless there is a good reason for not doing so! You can see why this order occurs, because we have a gross requirement for seats of 135 in week 3, why - because the Planned Order Release for the chair called for 135 chairs in week 3, and we need the seats to be available when we release the order to make the chairs. Other planned order releases for seats are 90 in week 3 and 60 in week 4.

Back

11-12-2000	Overdue	Week 1	Week 2	Week 3	Week 4
Item: Back		LT = 2	SS = 0	LS = LFL	UM = Each
Gross Requirement	0	0	0	135	0
Scheduled Receipt	0	10	0	0	0
Projected On Hand	40	50	50	0	0
Projected Net Requirement	0	0	0	85	0
Planned Order Receipt	0	0	0	85	0
Planned Order Release	0	85	0	90	60

Week 5	Week 6	Week 7	Week 8	Total
ABC = B	Source =	Type =	A. Demand = 0	@Cost = 0
90	60	0	0	285
0	0	0	0	10
0	0	0	0	
90	60	0	0	235
90	60	0	0	235
0	0	0	0	235

From this output we can see that we are advised to place an order for 85 backs in week 1. This advice should be followed unless there is a good reason for not doing so! You can see why this order occurs, because we have a gross requirement for backs of 135 in weeks 3, why - because the Planned Order Release for the chair called for 135 chairs in week 3, and we need the backs available to make the chairs. We already have 40 backs on-hand, another 10 projected to be received in week 1 so we only need to order 13510 – 40 = 85 new backs to have 135 backs available for the chairs in week 3.

Legs

11-12-2000	Overdue	Week 1	Week 2	Week 3	Week 4
Item: Legs		LT = 1	SS = 0	LS = LFL	UM = Each
Gross Requirement	0	0	0	540	0
Scheduled Receipt	0	0	0	0	0
Projected On Hand	80	80	80	0	0
Projected Net Requirement	0	0	0	460	0
Planned Order Receipt	0	0	0	460	0
Planned Order Release	0	0	460	0	360

Week 5	Week 6	Week 7	Week 8	Total
ABC = C	Source =	Type =	A. Demand = 0	@Cost = 0
360	240	0	0	1,140
0	0	0	0	0
0	0	0	0	
360	240	0	0	1,060
360	240	0	0	1,060
240	0	0	0	1,060

From this output we can see that we are advised to place an order for 460 legs in week 2. This advice should be followed *unless* there is a good reason for not doing so! You can see why this order occurs, because we have a gross requirement for legs of 540 in week 3, why - because the Planned Order Release for the chair called for 135 chairs in week 3, and we need the legs (4 per chair) available to make the chairs. Since we already have 80 legs on-hand we only need 540 – 80 = 460 new legs ordered.

Using the package we can get a summary of all actions (Planned Order Releases) as below.

11-12-2000	Item ID	Overdue	Week 1	Week 2	Week 3	Week 4	Week 5	Week 6	Week 7	Week 8	Total
1	Chair	0	0	0	135	0	90	60	0	0	285
2	Seat	0	25	0	90	60	0	0	0	0	175
3	Back	0	85	0	90	60	0	0	0	0	235
4	Legs	0	0	460	0	360	240	0	0	0	1,060

Sensitivity

It is often important to conduct sensitivity (what-if) analysis in MRP. Considering the example above, for instance, the only actions we need to take *now* relate to the orders suggested for the current period. All other actions relate to orders that we may (or may not) place at future points in time. These can be reviewed once the future arrives in the light of new information. Hence the orders suggested for the current period (for seats and backs in this case) are the ones to concentrate on with respect to sensitivity analysis.

So, for example, suppose the demand for the chair is more than expected in week 8. Does this impact upon the actions we must take now or not? If it does then we perhaps need to review the suggested order quantities for seats and backs accordingly.

To illustrate this suppose we change the demand for chairs in week 8 from 60 to 80, *i.e.* increase it. Changing the package data and re-exploding we get the action (planned order releases) list as compared with the previous action list of

11-12-2000	Item ID	Overdue	Week 1	Week 2	Week 3	Week 4	Week 5	Week 6	Week 7	Week 8	Total
1	Chair	0	0	0	135	0	90	80	0	0	305
2	Seat	0	25	0	90	80	0	0	0	0	195
3	Back	0	85	0	90	80	0	0	0	0	255
4	Legs	0	0	460	0	360	320	0	0	0	1,140

11-12-2000	Item ID	Overdue	Week 1	Week 2	Week 3	Week 4	Week 5	Week 6	Week 7	Week 8	Total
1	Chair	0	0	0	135	0	90	60	0	0	285
2	Seat	0	25	0	90	60	0	0	0	0	175
3	Back	0	85	0	90	60	0	0	0	0	235
4	Legs	0	0	460	0	360	240	0	0	0	1,160

Here it can be seen that the actions are identical until week 4. At that time we need to begin ordering more items to fulfill the projected increase in demand for chairs.

Extensions

To extend our example suppose that each leg is made up from two components (X and Y). Two units of X and 3 units of Y are needed for one leg and the lead time is 1 week. Then our BOM is:

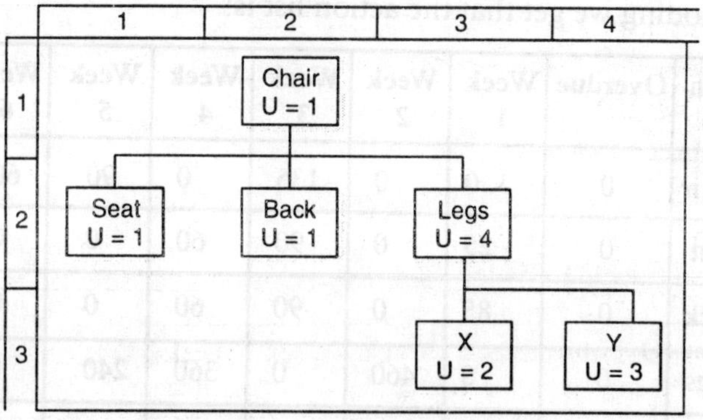

No.	Item ID	ABC Class	Source Code	Material Type	Unit Measure	Lead Time	Lot Szie
1	Chair	A			Each	2	LFL
2	Seat	B			Each	2	LFL
3	Back	B			Each	2	LFL
4	Legs	C			Each	1	LFL
5	X				Each	1	LFL
6	Y				Each	1	LFL

Item ID	Component ID/Usage	Component ID/Usage	Component ID/Usage
Chair	Seat	Back	Legs/4
Seat			
Back			
Legs	X/2	Y/3	
X			
Y			

Exploding we get that the action list is:

11-12-2000	Item ID	Overdue	Week 1	Week 2	Week 3	Week 4	Week 5	Week 6	Week 7	Week 8	Total
1	Chair	0	0	0	135	0	90	60	0	0	285
2	Seat	0	25	0	90	60	0	0	0	0	175
3	Back	0	85	0	90	60	0	0	0	0	235
4	Legs	0	0	460	0	360	240	0	0	0	1,060
5	X	0	920	0	720	480	0	0	0	0	2,120
6	Y	0	1,380	0	1,080	720	0	0	0	0	3,180

MRP II

MRP II is (essentially) MRP but with more added. Typically an MRP II package will include features such as cost information, management reports and easy "what-if" analysis. It may also include *capacitt requirements planning* (essentially try and automatically include capacity restrictions in the planning process). To confuse things MRP II stands for *manufacturing resources planning,* signifying that we are concentrating upon the planning of the manufacturing resources (*e.g.* people, machines, storage), rather than limiting ourselves to the planning of the materials requirements.

As an example of the use of MRP II we have the following:

ICI Agrochemicals

This company looked at over 40 MRP II packages before choosing Control: Manufacturing from Cincom. The manufacturing operation at their Fernhurst plant is complex with as many as 4,500 finished products being processed from 13,500 different raw materials. Before the days of MRP II stock levels were high and forecasting was hit and misses. A £1 million MRP II installation has helped bring about some radical operational improvements since it was installed. A good example has been in inventory records and bill of materials accuracy. A BOM administrator was appointed to monitor data accuracy with a regular audit of a dozen BOM's. In addition, staff had to be trained to a very high detailed level, learning how to structure the BOM's to support customer service objectives, inventory objectives, order entry objectives and costing objectives. Data integrity increased from around 65% to 98% in just six months as a result.

MRP Packages

There are many MRP packages available.

JIT

Just-In-Time (JIT) and MRP are two different systems for controlling production. It is often said that:

- MRP = a 'Push' system
- JIT = a 'Pull' system

Really this is an incorrect analysis - MRP is a system based on fulfilling *predicted* usage in a set time period. This can be seen in the example considered above. We never stocked out.

JIT is a system based on *actual* usage - parts of the production system are "linked" together via the use of Kanban's as the system runs. It is this linkage that is the distinguishing difference between MRP and JIT - JIT is a dynamic linked system, MRP is not. This implies that JIT can be used when lead times are short; MRP is more appropriate when lead times are long. In addition MRP is much better suited for computerized implementation then JIT. Consider, for example, the large number of finished products (4,500) and raw materials (13,500) mentioned in the ICI example considered above. Do you relish the idea of controlling that factory via a JIT system, or would you prefer a computerized MRP system?

Just-in-time (JIT)

Just-In-Time (JIT) is easy to grasp conceptually, everything happens *just-in-time.* For example consider my journey to work this morning, I could have left my house, *just-in-time* to catch a bus to the train station, just-in-time to catch the train, *just-in-time* to arrive at my ofiice, just-in-time to pick up my lecture notes, *just-in-time* to walk into this lecture theatre to start the lecture. Conceptually there is no problem about this, however achieving it in practice is likely to be difficult!. So too in a manufacturing operation component parts could conceptually arrive *just-in-time* to be picked up by a worker and used. So we would at a stroke eliminate any inventory of parts, they would simply arrive *just-in-time*! Similarly we could produce finished goods *just-in-time* to be handed to a customer who wants them. So, at a conceptual extreme, JIT has no need for inventory or stock, either of raw materials or work in progress or finished goods.

Obviously any sensible person will appreciate that achieving the conceptual extreme outlined above might well be difficult, or impossible, or extremely expensive, in real-life. However that extreme does illustrate that, perhaps, we could move an existing system towards a system with more of a JIT element than it currently contains. For example, consider a manufacturing process - whilst we might not be able to have a JIT process in terms of handing finished goods to customers, so we would still need some inventory of finished goods, perhaps it might be possible to arrange raw material deliveries so that, for example, materials needed for one day's production arrive at the start of the day and are consumed during the day - effectively reducing/eliminating raw material inventory.

Review Questions

1. What is Network Analysis? Write its salient features.
2. Define the following terms:
   (*i*) Pert
   (*ii*) CPM.
3. How 20 key Project Manager actions are organized? Explain.
4. What is material requirement planning (MRP)? Define it with suitable example.
5. How MRP is a 'Push' system while JIT is a 'Pull' system? Explain it.

MODEL QUESTION PAPERS

MODEL QUESTION PAPER I

Time : 3 Hrs. *Max. Marks : 100*

Note : 1. *Answer any* **Five** *questions.*
2. *All Question carry equal marks.*
3. *Use of statistical table allowed.*

1. (*a*) Explain in detail various phases of project life cycle with reference to a civil construction project or a software project. **(12 Marks)**

   (*b*) List and explain the characteristic features of a project. **(8 Marks)**

2. (*a*) Prepare a project feasibility report for manufacturing of washing machine considering marketing and technical factors. **(12 Marks)**

   (*b*) What is phased planning ? List the steps involved in project planning ? **(8 Marks)**

3. (*a*) What are the skills and abilities required for the project manager ? **(10 Marks)**

   (*b*) Explain in detail 3R's of contracting. **(10 Marks)**

4. (*a*) What do you understand by resource allocation ? Briefly explain the methods of resource allocation. **(6 Marks)**

(*b*) Ten men are available for a project, the details of which are given below :

Activity	*Duration (days)*	*No. of Men required*
1—2	1	07
2—3	2	03
2—4	3	05
2—5	2	01
3—6	2	02
4—5	1	09
5—6	3	10
3—4	3	06

(*i*) Draw the network and determine the normal length of the proejct.

(*ii*) Draw the time scaled version diagram of the network.

(*iii*) Prepare resource accumulation diagram.

(*iv*) Determine optimum schedule by resource levelling technique.

(14 Marks)

5. (*a*) In what ways does a CPM network differ from a PERT ? List out at least six differences. **(6 Marks)**

(*b*) A small project is composed of the following activities whose time estimates are given below : (Duration in weeks)

Activity	*Optimistic time* t_o	*Most likely* t_m	*Persimistic time* t_p
1—2	1	1	07
1—3	1	4	07
1—4	2	2	08
2—5	1	1	01
3—5	2	5	14
4—6	2	5	08
5—6	3	6	15

(*i*) Draw the network and find the critical path. **(5 Marks)**

(*ii*) Find the expected duration and variance for each activity. **(4 Marks)**

(*iii*) Calculate the standard deviation of the project. **(2 Marks)**

(*iv*) What is the probability that the project will be completed in 21 weeks ?

(3 Marks)

6. (*a*) Explain briefly
   (*i*) Project Direction and (*ii*) Project Co-ordination. **(6 + 6 Marks)**

   (*b*) Discuss the importance of communication in a project management. **(8 Marks)**

7. (*a*) Explain in brief how a construction management company (CM company) helps in project execution and management. **(10 Marks)**

   (*b*) Describe briefly with a neat sketch the project management environment in India. **(10 Marks)**

8. Write short notes on :
   (*a*) Work breakdown structure
   (*b*) Task force organisation
   (*c*) Gantt charts
   (*d*) Performance indicators. **(4 × 5 = 20 Marks)**

MODEL QUESTION PAPER II

Time: 3 Hrs.] *[Max. Marks: 100*

Note: 1. *Answer any* **Five** *questions.*
2. *All question carry equal marks.*
3. *Use of statistical table allowed.*

1. (*a*) Define a project and discuss the various categories of projects. **(10 Marks)**

(*b*) Discuss the roles and responsibilities of a project manager. **(10 Marks)**

2. (*a*) What is a feasibility report? Explain the various aspects of this report. **(10 Marks)**

(*b*) Briefly explain the different types of organization structures used in project management. **(10 Marks)**

3. (*a*) Discuss the 3 'R's of contracting. **(10 Marks)**

(*b*) Briefly explain the types of estimates drawn up during the life cycle of a project. **(10 Marks)**

4. (*a*) What is a 'work breakdown structure'? Explain with a suitable example. **(10 Marks)**

(*b*) Briefly explain different scheduling techniques. **(10 Marks)**

5. (*a*) What is a Gantt chart? Explain how it could be used in project control. **(10 Marks)**

(*b*) Discuss the various performance indicators of a project. **(10 Marks)**

6. (*a*) Define the terms:

(*i*) Slack (*ii*) Free float (*iii*) Independent float. **(6 Marks)**

(*b*) A project has 9 activities for which the times are given:

Activity	*Time in weeks*
1—2	8
1—3	5
2—4	9
3—5	7
2—5	5
4—6	3
5—6	6
6—7	10
5—7	9

Drawup the network and determine the critical path. **(14 Marks)**

7. (*a*) Discuss the role and importance of communication in project management. **(10 Marks)**

   (*b*) Briefly outline the role of MIS in project control. **(10 Marks)**

8. Write short notes on:

   (*a*) Tools and techniques in project management

   (*b*) Evaluation of project profitability

   (*c*) Communication feedback in project management

   (*d*) Computerised project management. **(5 × 4 = 20 Marks)**

MODEL QUESTION PAPER III

Time: 3 Hrs.] [*Max. Marks: 100*

Note: 1. *Answer any* **Five** *questions.*
2. *All question carry equal marks.*
3. *Draw neat block diagrams wherever necessary.*

1. (*a*) Define a Project. How is a project conceived? **(05 Marks)**
(*b*) What are the main characteristic features of a project? **(05 Marks)**
(*c*) Discuss on the areas to be examined for a cement plant during the definition phase of project life cycle. **(10 Marks)**

2. (*a*) List four basic reasons for the requirement of project planning. **(04 Marks)**
(*b*) Draw the block diagram of a typical project planning and control system. **(06 Marks)**
(*c*) Mention the factors that could result as a part of Technical Study Evaluation of a process plant during feasibility report preparation. **(10 Marks)**

3. (*a*) What are the basic objectives of a project? Give an example. **(06 Marks)**
(*b*) List five different methods of cost estimation and indicate their probable errors. **(06 Marks)**
(*c*) Discuss on cost-benefit analysis. **(08 Marks)**

4. (*a*) What is meant by Authority of a Project Manager? Under what areas of project management is authority requried? **(10 Marks)**
(*b*) Discuss the advantages and disadvantages of line/staff organization, project organization and matrix organization. **(10 Marks)**

5. (*a*) Explain briefly about the common framework provided by the Work Breakdown Structure (WBS). Draw the WBS tree diagram for a new car design. **(10 Marks)**
(*b*) What are the different techniques for effective Time Management of a project ? Write a note on learning curve. **(10 Marks)**

6. (*a*) Name any five different tools available under scheduling, coordinating and project execution. **(05 Marks)**
(*b*) What benefits are derived by applying PERT/CPM in Project Management ? **(05 Marks)**

(*c*) Discuss on computerized Project Management. **(10 Marks)**

7. (*a*) Write the procedure of estimating expected time of an activity, slack time and critical path pertaining to PERT network. **(08 Marks)**

(*b*) Consider the following details for a small project t_o, and t_m indicate optimistic time, most likely time and pessimistic time respectively in days. Draw the arrows diagram. Calculate the slack time. Mark the critical path with a bold line. What is the expected time for 68, 95 and 99 per cent completion limits? **(12 Marks)**

Activity	1–2	1–4	1–3	2–6	2–4	3–4	3–5	4–6	4–7	4–5	5–7	6–7
t_o	1	4	4	2	1	2	7	4	6	1	2	6
t_p	3	6	6	4	3	4	15	6	14	3	4	14
t_m	2	5	5	3	2	3	8	5	10	2	3	10

8. (*a*) What type of communication system is essential in project Management? Mention basic points required for a communication-oriented action plan. **(10 Marks)**

(*b*) Explain briefly about Project Performance Indicators with reference to Time overrun, cost overrun and productivity. **(10 Marks)**

MODEL QUESTION PAPER IV

Time: 3 Hrs.] [*Max. Marks: 100*

Note: 1. *Answer any* **Five** *questions.*
2. *All question carry equal marks.*
3. *Use of statistical table allowed.*

1. (*a*) What is project management? Discuss various categories of projects. **(10 Marks)**

(*b*) Explain different project life cycle phases with examples. **(10 Marks)**

2. (*a*) What are the factors to be considered in preparing feasibility report? Explain in detail. **(10 Marks)**

(*b*) Explain in detail about financial arrangements for a project. **(10 Marks)**

3. (*a*) Write in detail about cost estimates for a project. **(10 Marks)**

(*b*) What are the methods used for economic viability of the project? Explain in detail. **(10 Marks)**

4. (*a*) Explain different types of project organizations with neat sketch. **(10 Marks)**

(*b*) Define: (*i*) Tender (*ii*) Contract. Briefly explain the procedure for tendering and selection of contractors. **(10 Marks)**

5. (*a*) Define scheduling. What are its objectives? Explain different schecduling strategies. **(10 Marks)**

(*b*) Discuss the importance of Gantt chart in scheduling. **(04 Marks)**

(*c*) Apply Johnson rule to the following problem and find the best sequence, total minimum elapsed time and idle time for each machine in the order ABC. **(06 Marks)**

Job		1	2	3	4	5
Processing	A	8	10	6	7	11
times of	B	5	6	2	3	4
machines	C	4	9	8	6	5

6. (*a*) Write the applications of PERT/CPM technique. **(6 Marks)**

(*b*) A small project is composed of the following activities whose time estimates are given below: (duration in weeks).

Activity	Optimistic time, t_0	Most likely time, t_m	Pessimistic time, t_p
1—2	01	01	07
1—3	01	04	07
1—4	02	02	08
2—5	01	01	01
3—5	02	05	14
4—6	02	05	08
5—6	03	06	15

(*i*) Draw the network and find the critical path.

(*ii*) Find the expected duration and variance for each activity.

(*iii*) Calculate the standard deviation of the project.

(*iv*) What is the probability that the project will be completed in 21 weeks?

(14 Marks)

7. (*a*) What do you mean by direction, co-ordination and control of a project? Explain. **(10 Marks)**

   (*b*) Explain the steps involved in performance improvement of CM and DM Company. Explain in detail about CM and DM Company. **(10 Marks)**

8. Write short notes on any five of the following:

   (*a*) Tools and techinques in project management

   (*b*) Performance indicators

   (*c*) Work breakdown structure and project organization

   (*d*) CAT v/s RAT approach in scheduling

   (*e*) Roles and responsibilities of project manager and functional manager

   (*f*) Computerized project management. **(20 Marks)**

MODEL QUESTION PAPER V

Time: 3 Hrs.] [*Max. Marks: 100*

Note: 1. *Answer any* **Five** *questions.*

2. *All question carry equal marks.*

3. *Use of statistical table allowed.*

1. (*a*) Define project management. List the steps involved in it. **(08 Marks)**

(*b*) List the basic roles and responsibilities of project manager. **(08 Marks)**

(*c*) List the benefits of project management. **(04 Marks)**

2. (*a*) Brief on various types of cost estimates made for the project. **(10 Marks)**

(*b*) Brief on sources of financing for a project. **(10 Marks)**

3. (*a*) List the conditions for selection of a site for a project. **(06 Marks)**

(*b*) Brief on following methods of evaluation of project profitability:

(*i*) Pay back period (*ii*) ROI (*iii*) NPV. **(08 Marks)**

(*c*) List the task of project manager to achieve specific end objectives. **(06 Marks)**

4. (*a*) With table list the authority and responsibility of following types of organization:

(*i*) Matrix organization (*ii*) Task force. **(10 Marks)**

(*b*) List the factors to be considered while taking a decision on the number of contracts. **(05 Marks)**

(*c*) Brief on types of Reimbursement. **(05 Marks)**

5. (*a*) Brief on approaches for estimating the time duration of a project. **(08 Marks)**

(*b*) Brief on resource allocation methods. **(06 Marks)**

(c) Give an example for work breakdown structure—function oriented. **(06 Marks)**

6. (*a*) List the rules for Network Construction. **(05 Marks)**

(*b*) Consider the following problem involving activities from A to J:

Activity	*Immediate Predecessor (s)*	*Duration (months)*
A	—	1
B	A	4
C	A	2
D	A	2
E	D	3
F	D	3
G	E	2
H	F, G	1
I	C, H	3
J	B	2

(*i*) Construct the CPM network.

(*ii*) Determine the critical path.

(*iii*) Compute total floats and free floats for non-critical activities. **(15 Marks)**

7. (*a*) List the areas that need directions by the project manager. **(05 Marks)**

(*b*) List the actions to be taken for effective communications in a project. **(08 Marks)**

(*c*) With a sketch brief on how a project manager interacts in a DM company. **(07 Marks)**

8. Write short notes on the following:

(*a*) Computerized project management.

(*b*) Gantt chart.

(*c*) Project control.

(*d*) Value as performance indicator. **(20 Marks)**

MODEL QUESTION PAPER VI

Time: 3 Hrs.] [*Max. Marks: 100*

Note: 1. *Answer any* **Five** *questions.*
2. *All question carry equal marks.*
3. *Use of statistical table allowed.*

1. (*a*) Define project and with the help of a diagram discuss briefly the various categories of a project. **(10 Marks)**

(*b*) Explain clearly the different phases of project life cycle. **(10 Marks)**

2. (*a*) List and explain different feasibility reports to be considered for a project proposal. **(10 Marks)**

(*b*) Discuss about the phases of project planning. **(10 Marks)**

3. (*a*) Explain the various types of estimates that are prepared during the life cycle of project. **(10 Marks)**

(*b*) Give the list of methods of evaluation of project profitability and explain any two of them. **(10 Marks)**

4. (*a*) Define tender and explain the tendering procedure in a project. **(10 Marks)**

(*b*) Discuss about three 'R's of contract. **(10 Marks)**

5. (*a*) What is meant by scheduling? Explain the various guidelines to be followed in scheduling. **(10 Marks)**

(*b*) Explain briefly forward and backward scheduling and differentiate between them. **(10 Marks)**

6. (*a*) Discuss how GANTT chart is used as an effective tool in project management. Mention its merits and demerits. **(10 Marks)**

(*b*) Explain the following:

(*i*) The importance of communication

(*ii*) The various methods used for improving communication in project management. **(10 Marks)**

7. (*a*) Differentiate between PERT and CPM network. **(05 Marks)**

(*b*) The following table gives the information about activities and their time estimate of a project.

Activity	t_o	t_m	t_p
1—2	1	1.5	5
2—3	1	2	3
2—4	1	3	5
3—5	3	4	5
4—5	2	3	4
4—6	3	5	7
5—7	4	5	6
6—7	6	7	8
7—8	2	4	6
7—9	5	6	8
8—10	1	2	3
9—10	3	5	7

(*i*) Draw the network

(*ii*) Find the critical path and project duration

(*iii*) At 95% probability, what is the project duration? **(15 Marks)**

8. Write short notes on:

(*a*) Work breakdown structure,

(*b*) Performance indicators used in project management.

(*c*) MIS in project control,

(*d*) Reasons for ineffective project control. **(20 Marks)**

MODEL QUESTION PAPER VII

Time: 3 Hrs.] *[Max. Marks: 100*

Note: 1. *Answer any* **Five** *questions.*
2. *All question carry equal marks.*
3. *Use of statistical table allowed.*

1. (*a*) Discuss the various categories of a project. give an example for each category. **(10 Marks)**

(*b*) Explain the roles and responsibility of a project manager under various heads. **(10 Marks)**

2. (*a*) What is known as feasibility report? Explain its contents in detail. **(10 Marks)**

(*b*) List and explain five different methods of Project cost estimation. **(10 Marks)**

3. (*a*) Explain the methods of assessing the economic viability of the project. **(10 Marks)**

(*b*) Briefly explain the following: (*i*) Capital cost (*ii*) Working capital (*iii*) Operating cost (*iv*) Sources of financing. **(10 Marks)**

4. (*a*) What is delegation? Discuss the issues relating to delegation in detail. **(10 Marks)**

(*b*) Define and explain the project manager's authority. **(10 Marks)**

5. (*a*) List and explain various approaches of estimating time duration of a project. **(10 Marks)**

(*b*) What is work breakdown structure? Discuss with example the different basis of developing a work breakdown structure for a project. **(10 Marks)**

6. (*a*) Group and explain the various tools and techniques for effective project management. **(10 Marks)**

(*b*) The activities of a project and their time estimates are given below:

Activity	*Estimated duration of activity (in weeks)*		
	t_o	t_m	t_p
1—2	2	5	8
1—4	4	19	28
1—5	5	11	17
2—3	3	9	27
2—6	3	6	15
3—6	2	5	14
4—6	3	6	15
5—7	1	4	7
5—8	2	5	14
6—8	6	12	30
7—8	2	5	8

(*i*) Draw the PERT network diagram.

(*ii*) Determine the mean time and standard deviation of each activity.

(*iii*) Determine the critical path and the standard deviation for the critical path.

(*iv*) Calculate the total slack for the activities. **(10 Marks)**

7. (*a*) What is meant by Network? Discuss various types of network used for project management. **(10 Marks)**

(*b*) Discuss management efforts schedule during project implementation phase with a figure. **(10 Marks)**

8. (*a*) What is a performance indicator of a project? Explain in detail. **(10 Marks)**

(*b*) Differentiate CM Company and DM Company with block diagrams. **(10 Marks)**